Selected Titles in This Series

MW00910955

16 **Harish-Chandra,** Admissible invariant distributions on reductive p-adic groups (with notes by Stephen DeBacker and Paul J. Sally, Jr.), 1999

15 **Andrew Mathas,** Iwahori-Hecke algebras and Schur algebras of the symmetric group, 1999

14 **Lars Kadison,** New examples of Frobenius extensions, 1999

13 **Yakov M. Eliashberg and William P. Thurston,** Confoliations, 1998

12 **I. G. Macdonald,** Symmetric functions and orthogonal polynomials, 1998

11 **Lars Gårding,** Some points of analysis and their history, 1997

10 **Victor Kac,** Vertex algebras for beginners, Second Edition, 1998

 9 **Stephen Gelbart,** Lectures on the Arthur-Selberg trace formula, 1996

 8 **Bernd Sturmfels,** Gröbner bases and convex polytopes, 1996

 7 **Andy R. Magid,** Lectures on differential Galois theory, 1994

 6 **Dusa McDuff and Dietmar Salamon,** J-holomorphic curves and quantum cohomology, 1994

 5 **V. I. Arnold,** Topological invariants of plane curves and caustics, 1994

 4 **David M. Goldschmidt,** Group characters, symmetric functions, and the Hecke algebra, 1993

 3 **A. N. Varchenko and P. I. Etingof,** Why the boundary of a round drop becomes a curve of order four, 1992

 2 **Fritz John,** Nonlinear wave equations, formation of singularities, 1990

 1 **Michael H. Freedman and Feng Luo,** Selected applications of geometry to low-dimensional topology, 1989

Admissible Invariant Distributions on Reductive *p*-adic Groups

University
LECTURE
Series

Volume 16

Admissible Invariant Distributions on Reductive p-adic Groups

Harish-Chandra

Notes by
Stephen DeBacker and Paul J. Sally, Jr.

American Mathematical Society
Providence, Rhode Island

1991 *Mathematics Subject Classification.* Primary 22E50, 22E35.

ABSTRACT. This book contains a faithful rendering of Harish-Chandra's lectures on admissible invariant distributions (originally delivered at the Institute for Advanced Study in the 1970s). For a p-adic field Ω, let G denote the set of Ω-rational points of a connected reductive Ω-group. Let $\mathfrak{g}$ denote the Lie algebra of G. The main purpose of these notes is to show that the character of an irreducible admissible representation of G is represented by a locally summable function on G. The proof of this result begins with a study of harmonic analysis on $\mathfrak{g}$. The key result here is that the Fourier transform of a G-invariant distribution (with compactly generated support) on $\mathfrak{g}$ is itself represented by a locally summable function on $\mathfrak{g}$. Moreover, the Fourier transform of such a distribution has an asymptotic expansion about any semisimple point of $\mathfrak{g}$. This result, and most of the work on $\mathfrak{g}$, depends heavily on Howe's Theorem (for $\mathfrak{g}$) for which a proof is provided. Finally, these results are transferred to G via Howe's "Kirillov theory." These notes are intended for advanced graduate students and mathematicians working in this area.

Library of Congress Cataloging-in-Publication Data

Harish-Chandra.
   Admissible invariant distributions on reductive p-adic groups / Harish-Chandra ; notes by Stephen DeBacker and Paul J. Sally, Jr.
       p. cm. — (University lecture series, ISSN 1047-3998 ; v. 16)
   Includes bibliographical references and index.
   ISBN 0-8218-2025-7 (alk. paper)
   1. p-adic groups. 2. Distribution (Probability theory). I. DeBacker, Stephen, 1968– .
II. Sally, Paul. III. Title. IV. Series: University lecture series (Providence, R.I.) ; 16.
QA174.2.H37   1999
512′.74—dc21
                                                            99-31012
                                                            CIP

Contents

Preface   ix

Introduction   1

Part I. Fourier transforms on the Lie algebra   5
   1.   The mapping $f \mapsto \phi_{\hat{f}}$ 5
   2.   Some results about neighborhoods of semisimple elements   13
   3.   Proof of Theorem 3.1   17
   4.   Some consequences of Theorem 3.1   30
   5.   Proof of Theorem 5.11   32
   6.   Application of the induction hypothesis 38
   7.   Reformulation of the problem and completion of the proof   42
   8.   Some results on Shalika's germs   48
   9.   Proof of Theorem 9.6   51

Part II. An extension and proof of Howe's Theorem   55
   10. Some special subsets of $\mathfrak{g}$ 55
   11. An extension of Howe's Theorem   58
   12. First step in the proof of Howe's Theorem 62
   13. Completion of the proof of Howe's Theorem 63

Part III. Theory on the group   71
   14. Representations of compact groups   71
   15. Admissible distributions   74
   16. Statement of the main results   74
   17. Recapitulation of Howe's theory   75
   18. Application to admissible invariant distributions   76
   19. First step of the reduction from G to M 79
   20. Second step   82
   21. Completion of the proof   84
   22. Formal degree of a supercuspidal representation   86

Bibliography   91

List of Symbols   95

Index   97

Preface

Harish-Chandra first presented these notes on admissible distributions in lectures at the Institute for Advanced Study during 1973. In this preface, we provide a brief guide to the content of Harish-Chandra's notes and discuss the advances in this area of mathematics since these lectures were delivered. Of course, any such discussion will necessarily overlap Harish-Chandra's own introductory remarks (which begin on page 1).

A sketch of this material was published by Harish-Chandra in his Queen's notes [17]. Every statement in Harish-Chandra's Queen's notes also occurs here. Therefore, when we make a statement which occurs as an enumerated statement in the Queen's notes, we provide in parentheses the statement number appearing there (see, for example, the statement of Theorem 5.11).

A number of years ago, Harish-Chandra asked one of us (Sally) to produce a detailed version of his Queen's notes based on his own lecture notes. As was his custom, Harish-Chandra produced several versions of his lecture notes. We have made only minor changes to these, and most of these changes were with respect to the ordering. The two of us (DeBacker and Sally) carefully worked through Harish-Chandra's notes, and the version included here was typed by DeBacker. We take full responsibility for any errors.

The main results. Without further comment we adopt the terminology used by Harish-Chandra in [20].

Let Ω be a p-adic field of characteristic zero with ring of integers R. Let G be the group of Ω-rational points of a connected, reductive Ω-group. The group G, with its usual topology, is a locally compact, totally disconnected, unimodular group. In particular, it has a neighborhood basis of the identity consisting of compact open subgroups. Let dx denote the Haar measure on G and let G' denote the set of regular elements in G.

A complex representation (π, V) of G is *smooth* if, for each $v \in V$, there is an open subgroup K_v of G which fixes v (i.e., $\pi(k)v = v$ for all $k \in K_v$). The representation (π, V) is *admissible* if

(1) π is smooth, and
(2) for every compact open subgroup K of G, the space of K-fixed vectors has finite dimension.

Every irreducible and smooth representation is admissible [**29**]. Let (π, V) be an irreducible smooth representation of G. Denote by $C_c^\infty(G)$ the space of locally constant, compactly supported, complex-valued functions on G. For $f \in C_c^\infty(G)$, the operator

$$\pi(f) = \int_G f(x) \cdot \pi(x)\, dx$$

is an operator of finite rank. Consequently, it makes sense to define the *distribution character* of π by

$$\Theta_\pi(f) = \operatorname{tr} \pi(f)$$

for all $f \in C_c^\infty(G)$.

Motivated by the case of real reductive groups, we may ask if there exists a locally summable function F_π on G which is locally constant on G' such that

$$\Theta_\pi(f) = \int_G f(x) \cdot F_\pi(x)\, dx$$

for all $f \in C_c^\infty(G)$. It is the main purpose of these notes to provide an affirmative answer to this question. If F is an arbitrary nonarchimedean local field, then for the group $\mathrm{GL}_n(F)$ this result was established in the "tame" case by Rodier [**59**] and in the remaining cases by Lemaire [**39**]. In general, for the F-rational points of a connected reductive group defined over F, the most we can say is that the distribution character of an irreducible smooth representation is represented by a locally constant function on the set of regular elements [**22**] (see also Howe [**25**]).

One of the major results of these notes is a description of the behavior of Θ_π near a semisimple point γ of G (see Theorem 16.2). This is accomplished by deriving an asymptotic expansion for Θ_π in a neighborhood of γ. When γ is the identity element in G, we refer to this asymptotic expansion as the *local character expansion* of π. We need some definitions and notation before describing the local character expansion.

Let $\mathfrak{g}$ denote the Lie algebra of G. Let $C_c^\infty(\mathfrak{g})$ denote the space of complex-valued, locally constant, compactly supported functions on $\mathfrak{g}$. Let B be an Ω-valued, non-degenerate, symmetric, G-invariant bilinear form on $\mathfrak{g}$. Fix a non-trivial additive character χ on Ω. Let dX denote the Haar measure on the additive group of $\mathfrak{g}$ and, for $f \in C_c^\infty(\mathfrak{g})$, set

$$\hat{f}(Y) = \int_{\mathfrak{g}} f(X) \cdot \chi\big(B(X,Y)\big)\, dX$$

for $Y \in \mathfrak{g}$. The map $f \mapsto \hat{f}$ is a linear bijection of $C_c^\infty(\mathfrak{g})$ onto itself. If T is a distribution on $\mathfrak{g}$ (i.e., a linear functional on $C_c^\infty(\mathfrak{g})$), we define the Fourier transform $\widehat{T}$ of T by

$$\widehat{T}(f) = T(\hat{f})$$

for $f \in C_c^\infty(\mathfrak{g})$.

If $\mathcal{O}$ is a G-orbit in $\mathfrak{g}$ (under the adjoint action), then $\mathcal{O}$ carries a G-invariant measure which we denote by $\mu_{\mathcal{O}}$ [55]. It will follow from Theorem 4.4 that the Fourier transform of the distribution

$$f \mapsto \mu_{\mathcal{O}}(f)$$

for $f \in C_c^{\infty}(\mathfrak{g})$ is represented by a locally summable function on $\mathfrak{g}$ which is locally constant on $\mathfrak{g}'$, the set of regular elements of $\mathfrak{g}$. We denote this function by $\widehat{\mu_{\mathcal{O}}}$.

Since Ω has characteristic zero, the set of nilpotent orbits, which we denote by $\mathcal{O}(0)$, has finite cardinality. We can now state the local character expansion (see Theorem 16.2):

THEOREM. *Let π be an irreducible smooth representation of G. We can choose complex numbers $c_{\mathcal{O}}(\pi)$, indexed by $\mathcal{O} \in \mathcal{O}(0)$, such that*

$$\Theta_{\pi}(\exp Y) = \sum_{\mathcal{O} \in \mathcal{O}(0)} c_{\mathcal{O}}(\pi) \cdot \widehat{\mu_{\mathcal{O}}}(Y)$$

for all $Y \in \mathfrak{g}'$ sufficiently near zero.

This remarkable theorem, which was first proved by Howe [23] for the general linear group, is a qualitative result that leaves many unresolved quantitative questions. For example, almost no results exist about the quantitative nature of the $c_{\mathcal{O}}$s and the $\widehat{\mu_{\mathcal{O}}}$s. Moreover, outside of some stunning work of Waldspurger [72, 73] and a conjecture of Hales, Moy, and Prasad [43] we have only limited information about the precise range in which the equality holds.

Quantitatively, this is what we know about the $c_{\mathcal{O}}$s and $\widehat{\mu_{\mathcal{O}}}$s. For the general linear group, Howe [23] observed that the functions $\widehat{\mu_{\mathcal{O}}}$ have a very nice form (see also [41]) and showed that $c_{\mathcal{O}}(\pi)$ is an integer for every irreducible supercuspidal representation π and every nilpotent orbit $\mathcal{O}$. By using results of Kazhdan [31], Assem [1] determined the functions $\widehat{\mu_{\mathcal{O}}}$ for $\mathrm{SL}_{\ell}(\Omega)$ with ℓ a prime. Finally, by using a result later proved in general by Huntsinger [27], DeBacker and Sally [8] and Murnaghan [46] evaluated an integral formula to obtain values for the $\widehat{\mu_{\mathcal{O}}}$s in the cases $\mathrm{SL}_2(\Omega)$ and $\mathrm{GSp}_4(\Omega)$.

In Theorem 22.3 Harish-Chandra derives a formula for the leading term $c_0(\pi)$ in the local character expansion of an irreducible supercuspidal representation π of G. Strengthening a conjecture of Shalika [66], Harish-Chandra conjectures that this formula ought to hold for all irreducible discrete series representations of G. Rogawski proved this in [61]. Moreover, Huntsinger [28] used some work of Kazhdan [30] to show that for an irreducible tempered representation π, $c_0(\pi)$ is zero if and only if π is not a discrete series representation.

At the other extreme, Rodier [60] showed (for split G) that an irreducible admissible representation π has a Whittaker model if and only if there is a regular nilpotent orbit $\mathcal{O}$ such that $c_{\mathcal{O}}(\pi)$ is not zero. Mœglin and Waldspurger [41] refined this result. They showed that if $\mathcal{O}$ is maximal among those nilpotent orbits for which $c_{\mathcal{O}}(\pi)$ is nonzero, then the value of $c_{\mathcal{O}}(\pi)$ is related to the dimension of a degenerate or generalized Whittaker model. There have been many applications of these results. For classical groups Mœglin [40] showed that if $\mathcal{O}$ is maximal among those nilpotent orbits for which $c_{\mathcal{O}}(\pi)$ is nonzero, then the orbit $\mathcal{O}$ is special. Savin [65] showed that, for the representations constructed by Kazhdan and Savin in [32], the local character expansion involves only the trivial orbit and the minimal nilpotent orbits. A representation with this property is called a minimal representation. This work was extended by Rumelhart [63] and Torasso [69]. A version of Rodier's result for covering groups of $\mathrm{GL}_n(\Omega)$ is provided and used by Flicker and Kazhdan in [10].

In general, the remaining $c_{\mathcal{O}}$s have been calculated explicitly in only a few cases, most notably in the work of Assem [1], Barbasch and Moy [2], and Murnaghan [45, 46, 49, 50, 51]. In [13] Hales showed that most of the basic objects of harmonic analysis—including characters and the $\widehat{\mu_{\mathcal{O}}}$s—are non-elementary. That is, at some point, their values can be calculated by counting points on hyperelliptic curves over finite fields. Perhaps this is why these objects have been so hard to quantify explicitly.

A guide to these notes. The Lie algebra $\mathfrak{g}$ is a vector space over Ω of finite dimension, and G operates on $\mathfrak{g}$ by the adjoint representation, denoted Ad. Let T be a distribution on $\mathfrak{g}$. Then, for $x \in G$, the distribution ^xT is defined by

$$^xT(f) = T(f^x)$$

for $f \in C_c^\infty(\mathfrak{g})$ where

$$f^x(X) = f(\mathrm{Ad}(x)X)$$

for $X \in \mathfrak{g}$. The distribution T is said to be G-invariant if $^xT = T$ for all $x \in G$. Let J denote the space of all G-invariant distributions on $\mathfrak{g}$.

For $\omega \subset \mathfrak{g}$, let $J(\omega)$ denote the space of all G-invariant distributions T such that the support of T is contained in the closure of $\mathrm{Ad}(G)\,\omega$. If L is a lattice in $\mathfrak{g}$ (i.e., a compact open R-submodule of $\mathfrak{g}$) and T is a distribution on $\mathfrak{g}$, let $j_L T$ denote the restriction of T to $C_c(\mathfrak{g}/L)$. The following theorem, which was first conjectured by Howe in [26], makes nearly everything in these notes possible.

THEOREM 12.1 (Theorem 2). *Let ω be a compact set in $\mathfrak{g}$ and L a lattice in $\mathfrak{g}$. Then*

$$\dim j_L J(\omega) < \infty.$$

Although Howe [23] proved Theorem 12.1 only for the general linear group, Harish-Chandra [17] attributes this theorem to him. Consequently, Theorem 12.1 is often referred to as Howe's Theorem in the literature. Although Theorem 12.1 is used throughout Part I, the most significant applications can be found in §1.1, §4, and §5. In Part II of these notes Harish-Chandra states and proves an extension of Howe's Theorem which is used in Part III, §21. For the general linear group,

Howe [23] was the first to prove this extension. Waldspurger [76] also proved this extension of Howe's Theorem in the context of weighted orbital integrals. Waldspurger's proof includes the situation under consideration in these notes.

The key to understanding elements of J is Theorem 3.1. This theorem says that the regular semisimple orbital integrals are dense in J (i.e., if $f \in C_c^\infty(\mathfrak{g})$ and $\mu_\mathcal{O}(f) = 0$ for all G-orbits $\mathcal{O}$ which are contained in $\mathfrak{g}'$, then $T(f) = 0$ for all $T \in J$). This result, combined with an understanding of $\widehat{\mu_\mathcal{O}}$ for a regular orbit $\mathcal{O}$ (§3) and Howe's Theorem, allows Harish-Chandra to prove that $\widehat{\mu_\mathcal{O}}$ is represented by a locally summable function on $\mathfrak{g}$ which is locally constant on $\mathfrak{g}'$ (Theorem 4.4). Waldspurger [76] showed that far from zero, the function $\widehat{\mu_\mathcal{O}}$ has a particularly nice form. Another application of Howe's Theorem and some understanding of the geometry of open and closed G-invariant neighborhoods of zero permits Harish-Chandra to write down an asymptotic expansion of $\widehat{T}$ for $T \in J(\omega)$ with ω compact. Finally, in §7 Harish-Chandra derives an explicit integral formula for the Fourier transform of a regular orbital integral. This formula lets him see that the function representing $|\eta|^{1/2} \cdot \widehat{\mu_\mathcal{O}}$ is locally bounded on $\mathfrak{g}$ (here η is the usual discriminant).

The techniques of §7 were extended by Huntsinger [27] to show that the function representing a compactly supported distribution on $\mathfrak{g}$ has an integral formula. Rader and Silberger [54] extended a result of Harish-Chandra [21] to show that the character of an irreducible discrete series representation has an integral formula which is remarkably similar to the integral formula for the Fourier transform of a regular orbital integral obtained in §7. Murnaghan [47, 48, 50, 51] showed that this is not an accident and that in some cases the character of a supercuspidal representation can be related to the Fourier transform of a regular orbital integral. Murnaghan's work was extended by Cunningham [6] and DeBacker [7].

Following Shalika [66], in §8 Harish-Chandra develops the theory of what have become known as Shalika germs. In [56, 57, 58] Repka explicitly computed the Shalika germs corresponding to the regular and subregular unipotent orbits of $\mathrm{GL}_n(\Omega)$ and $\mathrm{SL}_n(\Omega)$ on the set of regular elliptic elements. (Kim [33, 34, 35] and Kim and So [36] partially computed the regular and subregular Shalika germs for $\mathrm{Sp}_4(\Omega)$.) For regular unipotent orbits, these results were extended to all groups by Shelstad [67], and for subregular unipotent orbits, they were extended to all groups by Hales [15]. For $\mathrm{GL}_n(\Omega)$, Rogawski [62] stated and proved (in some cases) a conjecture about the values of the Shalika germs evaluated at certain elliptic elements. Rogawski's conjecture for the Shalika germs of $\mathrm{GL}_n(\Omega)$ was confirmed by Waldspurger in [74]. This work was used by Murnaghan and Repka [52] to investigate which Shalika germs contribute to expansions about singular elliptic elements. For $\mathrm{GL}_n(\Omega)$, Waldspurger [75] provided an algorithm for computing Shalika germs which significantly extended the results of his earlier paper [74]. Courtès [5] extended this work of Waldspurger's to the group $\mathrm{SL}_n(\Omega)$. A few groups have had their Shalkia germs nearly completely worked out: Sally and Shalika [64, 66] computed them for $SL_2(\Omega)$ on the elliptic set, Langlands and Shelstad [38] calculated most of them for $SU(3, \Omega)$, and Hales [14, 16] worked them out for $GSp_4(\Omega)$ and $Sp_4(\Omega)$. There are many interesting questions surrounding Shalika germs and the theory of endoscopy which are beyond the scope of this discussion (see [37] and [71]).

Finally, in Part III Harish-Chandra studies admissible distributions on G. The goal of this section is to provide a way to transfer the results of Part I and Part II, which were concerned with G-invariant distributions on $\mathfrak{g}$, to admissible distributions on G. Suppose that we are only concerned with the behavior of our distribution on G near the identity. The definition of an admissible distribution, which Harish-Chandra attributes to the work of Howe (see [20, §16], and [24, 25, 26]), combined with some results derived from Howe's "Kirillov theory" ([24, 25]), allows him to derive from an admissible distribution on G a distribution on $\mathfrak{g}$ which

(1) satisfies the hypothesis of the extension of Howe's Theorem and

(2) is related to the original distribution on G via the exponential map.

Harish-Chandra is then able to conclude that near the identity, Θ_π is represented by a locally summable function on G.

There have been some generalizations of these notes to different settings. In [4], Clozel showed that the main results of these notes hold for non-connected groups. Clozel's paper also includes almost all of Part III of these notes. In [11] and [12], Hakim extended the content of these notes to certain symmetric spaces. However, in general, not everything in these notes can be extended to symmetric spaces. Rader and Rallis [53] generalized some of what can be carried over and provided counterexamples for those results which cannot be extended to the symmetric space setting. For further analysis of the symmetric space situation, see the work of Bosman [3] and Flicker [9]. Finally, Vignéras [70] explored the situation for modular representations.

We thank J. Adler, J. Boller, R. Huntsinger, D. Joyner, and M.-F. Vignéras for their extensive corrections and comments on earlier versions of these notes. We also thank J. Hakim, T. Hales, R. Kottwitz, G. Muić, F. Murnaghan, G. Savin, and the reviewer for their comments on earlier drafts of this opening.

Stephen DeBacker

Paul J. Sally, Jr.

The University of Chicago, 1999.

Introduction

Let Ω be a p-adic field and G the group of all Ω-rational points of a connected reductive Ω-group [20]. Then G, with its usual topology, is a locally compact, totally disconnected, unimodular group. Let dx denote the Haar measure of G. We shall use the terminology of [20] without further comment.

Let π be an admissible and irreducible representation of G and Θ_π the character of π. Let G' be the set of all points $x \in G$ where $D_G(x) \neq 0$ [20, §15]. Then G' is an open, dense subset of G whose complement has measure zero, and it can be shown that there exists a locally constant function F_π on G' such that

$$(\dagger) \qquad\qquad \Theta_\pi(f) = \int_G f(x) \cdot F_\pi(x)\, dx$$

for all $f \in C_c^\infty(G')$. One would like to prove that the function F_π is locally summable on G and that relation $(\dagger)$ actually holds for all $f \in C_c^\infty(G)$. The main object of these notes is to prove these facts when Ω has characteristic zero.

We assume from now on that char $\Omega = 0$. We also assume that the residue field of Ω has q elements.

THEOREM 16.3 (Theorem 1). *Let Θ_π denote the character of an admissible and irreducible representation of G. Then Θ_π is a locally summable function on G which is locally constant on G'. Moreover, the function*

$$|D_G|^{1/2} \cdot \Theta_\pi$$

is locally bounded on G.

We shall see presently that it is possible to describe the behavior of Θ_π around singular points somewhat more precisely.

Let $\mathfrak{g}$ be the Lie algebra of G. Then $\mathfrak{g}$ is a vector space over Ω of finite dimension, and G operates on $\mathfrak{g}$ by the adjoint representation. If $x \in G$ and $X \in \mathfrak{g}$, put

$$x.X = X^x = \mathrm{Ad}(x)X.$$

(If no misunderstanding is possible, we shall write xX instead of $x.X$.) Let T be a distribution on $\mathfrak{g}$. Then ^xT is defined to be the distribution $f \mapsto T(f^x)$ where

$$f^x(X) = f(x.X)$$

for $X \in \mathfrak{g}$ and $f \in \mathcal{D} = C_c^\infty(\mathfrak{g})$. T is said to be G-invariant if $^xT = T$ for all $x \in G$. Let J denote the space of all G-invariant distributions on $\mathfrak{g}$.

For two subsets $S \subset G$ and $\omega \subset \mathfrak{g}$, put

$$\omega^S = \bigcup_{x \in S} \mathrm{Ad}(x)\omega,$$

and let $J(\omega)$ denote the space of all $T \in J$ such that $\mathrm{Supp}(T) \subset \mathrm{cl}\,\omega^G$.

Let R denote the ring of all (p-adic) integers in Ω. By a lattice in $\mathfrak{g}$ we mean a compact open R-submodule of $\mathfrak{g}$. If L is a lattice, we may regard $C_c(\mathfrak{g}/L)$ as a subspace of $\mathcal{D}$. For any distribution T on $\mathfrak{g}$, let $T_L = j_L T$ denote the restriction of T on $C_c(\mathfrak{g}/L)$. The following remarkable theorem of Howe [**23**, **26**] is crucial for our theory.

THEOREM 12.1 (Theorem 2 (Howe)). *Let ω be a compact set in $\mathfrak{g}$ and L a lattice in $\mathfrak{g}$. Then*

$$\dim j_L J(\omega) < \infty.$$

Fix a symmetric, nondegenerate, G-invariant bilinear form B on $\mathfrak{g}$ with values in Ω. Since G is reductive, such a B exists. Also fix a character $\chi \neq 1$ of the additive group of Ω. Let dX denote a Haar measure on the additive group of $\mathfrak{g}$ and define

$$\hat{f}(Y) = \int_{\mathfrak{g}} \chi\big(B(Y, X)\big) \cdot f(X)\, dX$$

for $Y \in \mathfrak{g}$ and $f \in \mathcal{D}$. Then $f \mapsto \hat{f}$ is a linear bijection of $\mathcal{D}$ onto itself. For any distribution T on $\mathfrak{g}$, define the distribution $\widehat{T}$ by

$$\widehat{T}(f) = T(\hat{f})$$

for $f \in \mathcal{D}$. $\widehat{T}$ is called the Fourier transform of T.

Let ℓ denote the (absolute) rank of $\mathfrak{g}$. For $X \in \mathfrak{g}$, let $\eta_{\mathfrak{g}}(X)$ denote the coefficient of t^ℓ in the polynomial

$$\det(t - \mathrm{ad}\,X)$$

where t is an indeterminate. Then $\eta = \eta_{\mathfrak{g}}$ is a polynomial function on $\mathfrak{g}$, and $\eta \neq 0$. Let $\mathfrak{g}'$ be the set of all points $X \in \mathfrak{g}$ such that $\eta(X) \neq 0$.

THEOREM 4.4 (Theorem 3). *Let ω be a compact subset of $\mathfrak{g}$ and T an element in $J(\omega)$. Then there exists a locally summable function F on $\mathfrak{g}$ with the following properties.*

(1) $\widehat{T}(f) = \int_{\mathfrak{g}} F(X) \cdot f(X)\, dX$ *for all $f \in \mathcal{D}$.*
(2) *F is locally constant on $\mathfrak{g}'$.*
(3) *$|\eta|^{1/2} \cdot F$ is locally bounded on $\mathfrak{g}$.*

By an orbit (or more precisely a G-orbit) in $\mathfrak{g}$, we mean a set of the form X^G where X is an element of $\mathfrak{g}$. Fix an orbit, $\mathcal{O}$, and a point $X_0 \in \mathcal{O}$. Let $C_G(X_0)$ denote the centralizer of X_0 in G. Then $C_G(X_0)$ is unimodular, and therefore the homogeneous space $G/C_G(X_0)$ has an invariant measure dx^* which is unique up to a constant factor. By a theorem of Deligne and Rao [55], the integral

$$\mu_{\mathcal{O}}(f) = \int_{G/C_G(X_0)} f(x.X_0)\, dx^*$$

is well defined for $f \in C_c(\mathfrak{g})$. Hence $\mu_{\mathcal{O}}$ is a positive measure on $\mathfrak{g}$ which is uniquely determined by the orbit $\mathcal{O}$ up to a constant factor. It follows from Theorem 4.4 (Theorem 3) that $\widehat{\mu_{\mathcal{O}}}$ is a function.

Let $M_n(\Omega)$ denote the space of all $n \times n$ matrices with coefficients in Ω. Then one can speak of semisimple, unipotent, or nilpotent elements of $M_n(\Omega)$. Since, for a suitable n, both G and $\mathfrak{g}$ are subsets of $M_n(\Omega)$, such terms are applicable to their elements also. Let $\mathcal{N}$ be the set of all nilpotent elements in $\mathfrak{g}$. Then $\mathcal{N}$ is the union of a finite number of G-orbits which are called the nilpotent orbits.

By a G-domain in $\mathfrak{g}$ we mean a G-invariant subset of $\mathfrak{g}$ which is both open and closed. Let[1] $\mathcal{O}(0)$ denote the set of all nilpotent G-orbits in $\mathfrak{g}$.

THEOREM 5.11 (Theorem 4). *Let ω be a compact subset of $\mathfrak{g}$. Then there exists a G-domain D containing zero with the following property. For every $T \in J(\omega)$, we can choose complex numbers $c_{\mathcal{O}}(T)$ such that*

$$\widehat{T} = \sum_{\mathcal{O} \in \mathcal{O}(0)} c_{\mathcal{O}}(T) \cdot \widehat{\mu_{\mathcal{O}}}$$

on D. Moreover, if V is any neighborhood of zero in $\mathfrak{g}$, the functions $\widehat{\mu_{\mathcal{O}}}$, indexed by $\mathcal{O} \in \mathcal{O}(0)$, are linearly independent on $V \cap \mathfrak{g}'$.

We can now describe the behavior of a character around singular points by means of Theorem 5.11 (Theorem 4). Fix a semisimple point γ in G, and let M and $\mathfrak{m}$ denote the centralizers of γ in G and $\mathfrak{g}$, respectively. Define Θ_π as above (in Theorem 16.3 (Theorem 1)).

THEOREM 16.2 (Theorems 5 and 20). *We can choose unique complex numbers $c_\xi(\pi)$ such that*

$$\Theta_\pi(\gamma \exp Y) = \sum_\xi c_\xi(\pi) \cdot \widehat{\nu_\xi}(Y)$$

for all $Y \in \mathfrak{m}$ sufficiently near zero. Here ξ runs over all nilpotent M-orbits in $\mathfrak{m}$, ν_ξ is the M-invariant measure on $\mathfrak{m}$ corresponding to ξ, and $\widehat{\nu_\xi}$ is the Fourier transform of ν_ξ on $\mathfrak{m}$.

Now consider the special case when γ is the identity in G. Then $M = G$, $\mathfrak{m} = \mathfrak{g}$, and $\{0\}$ is a nilpotent G-orbit in $\mathfrak{g}$. Let $c_0(\pi)$ denote the coefficient corresponding to this orbit. Assume further that π is supercuspidal and unitary, and let $d(\pi)$ denote the formal degree of π.

THEOREM 22.3 (Theorem 6). *There exists a real number $c \neq 0$ such that*

$$c_0(\pi) = c \cdot d(\pi)$$

---

[1]An orbit $\mathcal{O}$ is nilpotent if and only if $\mathrm{cl}\,\mathcal{O}$ contains zero. This justifies our notation (see also §2).

for every irreducible, unitary, supercuspidal π.

It seems likely that this relation actually holds for all square-integrable π. This would imply that

$$c = (-1)^{\ell_0} \cdot d(\pi_0)^{-1}$$

where π_0 is the Steinberg representation and $\ell_0 = \dim(A_0/Z)$ in the notation of [**20**, §15].

THEOREM 22.6 (Theorem 7). *It is possible to normalize the Haar measure on G/Z in such a way that $d(\pi)$ is an integer for every irreducible, unitary, supercuspidal π.*

For $GL(n)$ this has been proven by Howe [**23**] for arbitrary characteristic.

This paper is divided into three parts. In Part I we concentrate on Fourier transforms on the Lie algebra. The main object here is to derive Theorem 4.4 (Theorem 3) and Theorem 5.11 (Theorem 4) from Howe's Theorem (Theorem 12.1), and our argument depends in an essential way on Theorem 3.1 and Theorem 7.7. Part II is devoted to an extension and proof of Theorem 12.1. The extension is necessary for the application of these results to the group. In Part III we recall the main points of Howe's "Kirillov theory" [**24, 25**] and introduce the concept of an admissible distribution. This concept, which was suggested by the work of Howe (see [**20**, §16] and [**24, 25, 26**]), is central to our method.

Some of these results have been announced in a brief note [**17**].

Part I. Fourier transforms on the Lie algebra

1. The mapping $f \mapsto \phi_{\hat{f}}$

Fix a Cartan subalgebra $\mathfrak{h}$ of $\mathfrak{g}$ and put $\mathfrak{h}' = \mathfrak{h} \cap \mathfrak{g}'$. Let A denote the split component of the corresponding Cartan subgroup of G. $\mathfrak{h}$ is called elliptic if A lies in the center of G.

For $f \in \mathcal{D} = C_c^{\infty}(\mathfrak{g})$ we define a function ϕ_f on $\mathfrak{h}'$ by

$$\phi_f(H) = |\eta(H)|^{1/2} \int_{G/A} f(x.H)\, dx^*$$

where dx^* is the invariant measure on G/A. Then ϕ_f is locally constant on $\mathfrak{h}'$. Sometimes it would be convenient to write $\phi_f^{\mathfrak{h}}$ or $\phi_f^{G/\mathfrak{h}}$ instead of ϕ_f.

Fix an element H_0 in $\mathfrak{h}'$.

THEOREM 1.1 (Theorem 8). *The distribution*

$$f \mapsto \phi_{\hat{f}}(H_0)$$

for $f \in \mathcal{D}$ is a locally summable function on $\mathfrak{g}$ which is locally constant on $\mathfrak{g}'$. Let F denote this function. Then $|\eta|^{1/2} \cdot F$ is locally bounded on $\mathfrak{g}$.

We shall prove this theorem by induction on $\dim G$. But first we need some preparation.

1.1. Some consequences of Howe's Theorem (Theorem 12.1).

LEMMA 1.2. *Fix a Cartan subalgebra $\mathfrak{h}$ of $\mathfrak{g}$ and fix $H \in \mathfrak{h}'$. Consider the distribution*

$$T_H \colon f \mapsto \phi_{\hat{f}}(H)$$

for $f \in \mathcal{D}$. Let ω and ω_0 be compact subsets of $\mathfrak{g}$ and $\mathfrak{h}$, respectively, with ω open. Let τ_H be the restriction of T_H on ω. Then the space spanned by the set of distributions $\{\tau_H \colon H \in \omega_0' = \omega_0 \cap \mathfrak{h}'\}$ is finite dimensional.

PROOF. Put

$$\sigma_H(f) = \phi_f(H)$$

for $H \in \mathfrak{h}'$. Then $\sigma_H \in J(\omega_0)$ for $H \in \omega_0'$. Since ω is compact, we can choose a lattice L in $\mathfrak{g}$ such that $\hat{f} \in C_c(\mathfrak{g}/L)$ for all $f \in C_c^{\infty}(\omega)$. Hence, for $f \in C_c^{\infty}(\omega)$ and $H \in \omega_0'$,

$$\tau_H(f) = \sigma_H(\hat{f}) = \sigma_{H,L}(\hat{f})$$

where $\sigma_{H,L} = j_L \sigma_H$. Since $\dim j_L J(\omega_0) < \infty$, our assertion follows. $\qquad \square$

COROLLARY 1.3. *Fix $H_0 \in \mathfrak{h}'$, and fix a compact open set ω in $\mathfrak{g}$. Then we can choose a neighborhood ω_0 of H_0 in $\mathfrak{h}'$ such that $\phi_{\hat{f}}$ is constant on ω_0 for all $f \in C_c^\infty(\omega)$.*

PROOF. Fix a compact neighborhood ω_1 of H_0 in $\mathfrak{h}'$. The distributions τ_H, indexed by $H \in \omega_1$, span a finite dimensional space. Choose $H_1, H_2, \ldots, H_r \in \omega_1$ such that the $\tau_i = \tau_{H_i}$ form a base for this space. Choose $f_j \in C_c^\infty(\omega)$ such that $\tau_i(f_j) = \delta_{ij}$. Now, if $H \in \omega_1$, then

$$\tau_H = \sum_{i=1}^r \tau_H(f_i) \cdot \tau_i = \sum_{i=1}^r \phi_{\hat{f}_i}(H) \cdot \tau_i.$$

Since $\phi_{\hat{f}_i}$ is a locally constant function on $\mathfrak{h}'$ for $1 \le i \le r$, there exists a neighborhood ω_0 of H_0 in ω_1 such that the functions $\phi_{\hat{f}_i}$ are constant on ω_0. So, for $f \in C_c^\infty(\omega)$ and $H \in \omega_0$,

$$\phi_{\hat{f}}(H) = \tau_H(f) = \sum_{i=1}^r \phi_{\hat{f}_i}(H) \cdot \tau_i(f) = \sum_{i=1}^r \phi_{\hat{f}_i}(H_0) \cdot \tau_i(f)$$
$$= \phi_{\hat{f}}(H_0). \quad \square$$

1.2. Parabolic descent. Suppose (P, A) is a p-pair in G and $P = MN$. Let $\bar{P} = M\bar{N}$ be the p-pair opposite (P, A). Then $\mathfrak{g} = \bar{\mathfrak{n}} + \mathfrak{m} + \mathfrak{n}$ where $\bar{\mathfrak{n}}$, $\mathfrak{m}$, and $\mathfrak{n}$ are the Lie algebras of $\bar{N}$, M, and N, respectively. Further, the sum is direct. Let dX, $d\bar{Z}$, dY, and dZ denote the Haar measures on the additive groups of $\mathfrak{g}$, $\bar{\mathfrak{n}}$, $\mathfrak{m}$, and $\mathfrak{n}$, respectively. We normalize them in such a way that

$$dX = d\bar{Z}\, dY\, dZ.$$

For $f \in \mathcal{D}$, define $g_f \in C_c^\infty(\mathfrak{m})$ by

$$g_f(Y) = \int_{\mathfrak{n}} f(Y + Z)\, dZ.$$

LEMMA 1.4. *We have*

$$g_{\hat{f}} = (g_f)\hat{\ }.$$

Here $(g_f)\hat{\ }$ is the Fourier transform of g_f on $\mathfrak{m}$.

PROOF. Let $X_i = \bar{Z}_i + Y_i + Z_i$ where $\bar{Z}_i \in \bar{\mathfrak{n}}$, $Y_i \in \mathfrak{m}$, and $Z_i \in \mathfrak{n}$. Then

$$B(X_1, X_2) = B(\bar{Z}_1, Z_2) + B(Y_1, Y_2) + B(Z_1, \bar{Z}_2).$$

Hence

$$\hat{f}(Y_1 + Z_1) = \int_{\bar{\mathfrak{n}}+\mathfrak{m}+\mathfrak{n}} \chi\big(B(Y_1, Y_2) + B(Z_1, \bar{Z}_2)\big) \cdot f(Z_2 + Y_2 + \bar{Z}_2)\, dZ_2\, dY_2\, d\bar{Z}_2$$
$$= \int_{\bar{\mathfrak{n}}} \chi\big(B(Z_1, \bar{Z}_2)\big)\, d\bar{Z}_2 \int_{\mathfrak{n}+\mathfrak{m}} \chi\big(B(Y_1, Y_2)\big) \cdot f(Z_2 + Y_2 + \bar{Z}_2)\, dZ_2\, dY_2.$$

Therefore

$$\int_{\mathfrak{n}} \hat{f}(Y_1 + Z_1)\, dZ_1 = \int_{\mathfrak{n}+\mathfrak{m}} \chi\big(B(Y_1, Y_2)\big) \cdot f(Z_2 + Y_2)\, dZ_2\, dY_2.$$

That is,

$$g_{\widehat{f}} = (g_f)^{\widehat{}}.$$

(If $X = \bar{Z} + Y + Z$ and $dX = d\bar{Z}\, dY\, dZ$, then dX is self dual with respect to B, $d\bar{Z}$ and dZ are dual to each other with respect to $B(\bar{Z}, Z)$, and dY is self dual with respect to restriction of B on $\mathfrak{m}$.)   □

Fix an open compact subgroup K of G of Bruhat-Tits [**21**, Theorem 5, p. 16]. Since K is a maximal special compact open subgroup, we have $G = KP$. Let dk be the normalized Haar measure on K. For $f \in \mathcal{D}$ and $X \in \mathfrak{g}$, put

$$\bar{f}(X) = \int_K f(k.X)\, dk$$

and define

$$f_P = g_{\bar{f}}$$

so that

$$f_P(Y) = \int_{\mathfrak{n}} \bar{f}(Y + Z)\, dZ$$

for $Y \in \mathfrak{m}$.

LEMMA 1.5. *If Γ is a Cartan subalgebra of $\mathfrak{m}$, then $\phi_f^{G/\Gamma} = \phi_{f_P}^{M/\Gamma}$.*

PROOF. Let A_Γ be the split component of C_Γ, where C_Γ is the Cartan subgroup of G corresponding to Γ. Then, for $\gamma \in \Gamma'$,

$$\phi_f^{G/\Gamma}(\gamma) = |\eta(\gamma)|^{1/2} \int_{G/A_\Gamma} f(x.\gamma)\, dx^*$$

where dx^* is the invariant measure on G/A_Γ.

Since $G = KNM$, the Haar measure dx on G may be written as $dx = dk\, dn\, dm$ where dk, dn, and dm are the Haar measures on K, N, and M, respectively. Fix a Haar measure da on A_Γ and normalize the invariant measures dx^* and dm^* on G/A_Γ and M/A_Γ, respectively, so that

$$dx = dx^*\, da \quad \text{and} \quad dm = dm^*\, da.$$

It is obvious that for $\gamma \in \Gamma'$

$$\phi_f^{G/\Gamma}(\gamma) = |\eta(\gamma)|^{1/2} \int_{M/A_\Gamma \times N} \bar{f}\big(n.(m.\gamma)\big)\, dm^*\, dn.$$

On the other hand, we have the following result.

LEMMA 1.6. *Let $g \in C_c^\infty(\mathfrak{m} + \mathfrak{n})$. Then, for $Y \in \mathfrak{m}$,*

$$|\det(\operatorname{ad} Y)_{\mathfrak{n}}| \int_N g(n.Y)\, dn = \int_{\mathfrak{n}} g(Y + Z)\, dZ$$

provided that $\det(\operatorname{ad} Y)_{\mathfrak{n}} \neq 0$.

Here we assume that dZ goes into dn under the exponential mapping. Over $\mathbb{R}$ this is done in [**19**, pp. 217–218]. The p-adic case can be handled in more or less the same way (see [**21**, Lemma 22, p. 58]).

For $Y \in \mathfrak{m}$, put

$$\eta_{\mathfrak{g}/\mathfrak{m}}(Y) = \det(\operatorname{ad} Y)_{\mathfrak{g}/\mathfrak{m}} = \det(\operatorname{ad} Y)_{\mathfrak{n}} \cdot \det(\operatorname{ad} Y)_{\bar{\mathfrak{n}}}.$$

Since B defines a nondegenerate, bilinear form on $\bar{\mathfrak{n}} \times \mathfrak{n}$, we may regard $\bar{\mathfrak{n}}$ as the space dual to $\mathfrak{n}$. Moreover, for $Z \in \mathfrak{n}$ and $\bar{Z} \in \bar{\mathfrak{n}}$,

$$B(\bar{Z}, [Y, Z]) = -B([Y, \bar{Z}], Z).$$

Therefore $(-\operatorname{ad} Y)_{\bar{\mathfrak{n}}}$ is the 'transpose' of $(\operatorname{ad} Y)_{\mathfrak{n}}$ and so

$$\left| \eta_{\mathfrak{g}/\mathfrak{m}}(Y) \right| = \left| \det(\operatorname{ad} Y)_{\mathfrak{n}} \right|^2.$$

Then, for $m \in M$,

$$\left| \eta_{\mathfrak{g}/\mathfrak{m}}(m.\gamma) \right| = \left| \eta_{\mathfrak{g}/\mathfrak{m}}(\gamma) \right| = \left| \det(\operatorname{ad} \gamma)_{\mathfrak{n}} \right|^2.$$

So

$$\left| \eta_{\mathfrak{g}/\mathfrak{m}}(m.\gamma) \right|^{1/2} = \left| \det(\operatorname{ad} \gamma)_{\mathfrak{n}} \right|.$$

Define the polynomial function $\eta_{\mathfrak{m}}$ on $\mathfrak{m}$ corresponding to η on $\mathfrak{g}$. It is clear that $\eta = \eta_{\mathfrak{m}} \cdot \eta_{\mathfrak{g}/\mathfrak{m}}$ on $\mathfrak{m}$. Hence

$$\phi_f^{G/\Gamma}(\gamma) = |\eta_{\mathfrak{m}}(\gamma)|^{1/2} \int_{M/A_\Gamma \times \mathfrak{n}} \bar{f}(m.\gamma + Z) \, dm^* \, dZ$$

$$= |\eta_{\mathfrak{m}}(\gamma)|^{1/2} \int_{M/A_\Gamma} f_P(m.\gamma) \, dm^* = \phi_{f_P}^{M/\Gamma}(\gamma). \quad \square$$

LEMMA 1.7. *Let $f \in \mathcal{D}$. Then $(\hat{f})_P = (f_P)\hat{\ }$ and therefore*

$$\phi_{\hat{f}}^{G/\Gamma} = \phi_{(f_P)\hat{\ }}^{M/\Gamma}$$

for any Cartan subalgebra Γ of $\mathfrak{m}$.

PROOF. The fact that $(\hat{f})_P = (f_P)\hat{\ }$ follows from Lemma 1.4 and the fact that $\hat{\bar{g}} = \bar{\hat{g}}$ for $g \in \mathcal{D}$. The rest is obvious. $\quad \square$

1.3. Weyl's integration formula and some consequences. Let $\mathfrak{h}_1$, $\mathfrak{h}_2$, ..., $\mathfrak{h}_r$ be a complete set of Cartan subalgebras of $\mathfrak{g}$ no two of which are conjugate under G. Let A_i be the split component of the Cartan subgroup corresponding to $\mathfrak{h}_i$. Fix an invariant measure $d_i x^*$ on G/A_i and a Haar measure dH_i on the additive group of $\mathfrak{h}_i$. The following result is well known.

LEMMA 1.8 (Weyl's Integration Formula). *There exist positive numbers c_i such that, for $f \in C_c(\mathfrak{g})$,*

$$\int_{\mathfrak{g}} f(X) \, dX = \sum_{i=1}^r c_i \int_{\mathfrak{h}_i} |\eta(H_i)| \, dH_i \int_{G/A_i} f(x.H_i) \, d_i x^*.$$

It follows from [**21**, Theorem 13, p. 64] that for any $f \in \mathcal{D}$ and any Cartan subalgebra Γ of $\mathfrak{g}$,

$$\sup_{\gamma \in \Gamma'} \left| \phi_f^\Gamma(\gamma) \right| < \infty.$$

COROLLARY 1.9. *Let θ be a G-invariant measurable function on $\mathfrak{g}$. Then the following two conditions on θ are equivalent.*

(1) *θ is a locally summable function on $\mathfrak{g}$.*

(2) *For every Cartan subalgebra Γ of $\mathfrak{g}$ and $f \in \mathcal{D}$*

$$\gamma \mapsto |\eta(\gamma)|^{1/2} \cdot \theta(\gamma) \cdot \phi_f^\Gamma(\gamma) \quad (\gamma \in \Gamma')$$

is an integrable function on Γ.

PROOF. This is immediate from Lemma 1.8. $\square$

COROLLARY 1.10. *There exists a number $\varepsilon > 0$ such that $|\eta|^{-1/2-\varepsilon}$ is locally summable on $\mathfrak{g}$.*

PROOF. This follows from Corollary 1.9 and [**21**, Lemma 44, p. 89]. $\square$

COROLLARY 1.11. *Define $\lambda(X)$ by $|\eta(X)| = q^{\lambda(X)}$. Then $|\eta|^{-1/2} \cdot (1 + |\lambda|)^r$ is locally summable on $\mathfrak{g}$ for any $r \geq 0$.*

PROOF. This is obvious given the previous corollary. $\square$

1.4. Reduction to the elliptic case. Let A_1 and A_2 be two special tori in G. We define $w(A_2|A_1)$ as usual and write $A_1 \prec A_2$ if $w(A_2|A_1) \neq \emptyset$. We also write $A_1 \asymp A_2$ if $A_1 \prec A_2$ and $A_1 \succ A_2$ (i.e., A_1 and A_2 are conjugate in G). For a Cartan subalgebra Γ of $\mathfrak{g}$, we denote by A_Γ the split component of the corresponding Cartan subgroup C_Γ of G.

Let (P, A) be a p-pair with $P = MN$, and let θ be a distribution on $\mathfrak{m}$. For $f \in \mathcal{D}$ put

$$\Theta(f) = \theta(f_P).$$

Then Θ is a distribution on $\mathfrak{g}$.

LEMMA 1.12. *Suppose θ is M-invariant. Then Θ is G-invariant.*

PROOF. Define a representation ρ of G on $\mathcal{D}$ by

$$\rho(x^{-1})f = f^x.$$

In order to prove that Θ is G-invariant, it is enough to verify that

$$\Theta\big(\rho(\alpha)f\big) = \Theta(f) \int_G \alpha(x) \, dx$$

for all $\alpha \in C_c^\infty(G)$. It is easy to deduce from the definitions of Θ and f_P that $^k\Theta = \Theta$ for $k \in K$ and therefore

$$\Theta\big(\rho(\alpha)f\big) = \Theta\big(\rho(\alpha_0)f\big)$$

where $\alpha_0(x) = \int_K \alpha(kx) \, dk$. Hence we may assume that $\alpha(kx) = \alpha(x)$ for $k \in K$.

Now $G = KNM$ and $dx = dk\, dn\, dm$ as above. Hence, if $Y \in \mathfrak{m}$,

$$
\big(\rho(\alpha)f\big)_P(Y) = \int_{G \times \mathfrak{n}} \alpha(x^{-1}) \cdot f\big(x.(Y + Z)\big)\, dx\, dZ
$$

$$
= \int_{\mathfrak{n} \times N \times M \times K} \alpha\big((knm)^{-1}\big) \cdot f\big(k.(Y^m + Z^m)\big)\, dk\, dm\, dn\, dZ.
$$

But

$$
d(Z^m) = \delta_P(m)dZ.
$$

Hence

$$
\big(\rho(\alpha)f\big)_P(Y) = \int_{\mathfrak{n} \times N \times M \times K} \alpha\big((knm)^{-1}\big) \cdot \delta_P(m^{-1}) \cdot f\big(k.(Y^m + Z)\big)\, dk\, dm\, dn\, dZ.
$$

So

$$
\Theta\big(\rho(\alpha)f\big) = \int_{N \times M \times K} \alpha\big((knm)^{-1}\big) \cdot \delta_P(m^{-1}) \cdot \theta(g_{\rho(k^{-1})f})\, dk\, dm\, dn.
$$

But

$$
\int_{N \times M} \alpha\big((knm)^{-1}\big) \cdot \delta_P(m^{-1})\, dm\, dn = \int_{N \times M} \alpha\big(mnk^{-1}\big) \cdot \delta_P(m)\, dm\, dn
$$

$$
= \int_G \alpha(xk^{-1})\, dx = \int_G \alpha(x)\, dx.
$$

Hence

$$
\Theta\big(\rho(\alpha)f\big) = \int_G \alpha(x)\, dx \cdot \int_K \theta(g_{\rho(k^{-1})f})\, dk
$$

$$
= \int_G \alpha(x)\, dx \cdot \theta(f_P) = \int_G \alpha(x)\, dx \cdot \Theta(f). \quad \square
$$

LEMMA 1.13. *Suppose θ is a locally summable, M-invariant function on $\mathfrak{m}$. Then Θ is a locally summable, G-invariant function on $\mathfrak{g}$. If Γ is a Cartan subalgebra of $\mathfrak{g}$, then, for $\gamma \in \Gamma'$,*

$$
|\eta(\gamma)|^{1/2} \cdot \Theta(\gamma) = \sum_{s \in w(A_\Gamma | A)} |\eta_{\mathfrak{m}}^s(\gamma)|^{1/2} \cdot \theta^s(\gamma).
$$

PROOF. Here $\theta^s(\gamma) = \theta(y^{-1}.\gamma)$ and $\eta_{\mathfrak{m}}^s(\gamma) = \eta_{\mathfrak{m}}(y^{-1}.\gamma)$ where y is a representative of s in G. Note that $y^{-1}.\Gamma \subset \mathfrak{m}$.

The proof here is parallel to the proof of [**68**, Lemma 4.7.6]. $\quad \square$

COROLLARY 1.14. *If $|\eta_{\mathfrak{m}}|^{1/2} \cdot \theta$ is locally bounded on $\mathfrak{m}$, then $|\eta|^{1/2} \cdot \Theta$ is locally bounded on $\mathfrak{g}$.*

PROOF. This is clear. $\quad \square$

Now we come to the proof of Theorem 1.1. Let $A = A_{\mathfrak{h}}$. Then A is a special torus. Now assume that $\mathfrak{h}$ is not elliptic. Then $A \neq Z$ where Z is the split component of G. Fix $P \in \mathcal{P}(A)$ with $P = MN$. Then, for $f \in \mathcal{D}$,

$$
\phi_{\hat{f}}^{G/\mathfrak{h}} = \phi_{(f_P)^{\widehat{}}}^{M/\mathfrak{h}}
$$

from Lemma 1.7. Since $A \neq Z$, we have $\dim M < \dim G$. Hence, by the induction hypothesis, there exists an M-invariant, locally summable function θ on $\mathfrak{m}$ such that

(1) θ is locally constant on $\mathfrak{m}' = \mathfrak{m} \cap \mathfrak{g}'$,
(2) $|\eta_{\mathfrak{m}}|^{1/2} \cdot \theta$ is locally bounded on $\mathfrak{m}$, and
(3) $\phi_{\hat{g}}^{M/\mathfrak{h}}(H_0) = \int_{\mathfrak{m}} g(Y) \cdot \theta(Y) \, dY$ for all $g \in C_c^\infty(\mathfrak{m})$.

Then the conditions required by Theorem 1.1 follow immediately from Lemma 1.13.

So in order to complete the proof of Theorem 1.1 we may assume that $\mathfrak{h}$ is elliptic.

1.5. Some distributions for the elliptic case. Let $\mathfrak{h}$ be an elliptic Cartan subalgebra of $\mathfrak{g}$. As usual, put

$$\mathfrak{g}_{\mathfrak{h}} = (\mathfrak{h}')^G.$$

Then $\mathfrak{g}_{\mathfrak{h}}$ is an open, G-invariant subset of $\mathfrak{g}$.

LEMMA 1.15. *Fix $g \in C_c^\infty(\mathfrak{g}_{\mathfrak{h}})$. Then, for $f \in \mathcal{D}$,*

$$\int_{G/Z} dx^* \left| \int_{\mathfrak{g}} f(X) \cdot \hat{g}(x.X) \, dX \right| < \infty.$$

Here dx^* is the Haar measure on G/Z.

PROOF. It follows from Lemma 1.8 that

$$\int_{\mathfrak{g}} f(X) \cdot \hat{g}(x.X) \, dX = \int_{\mathfrak{g}} \hat{f}(X) \cdot g(x.X) \, dX$$

$$= \int_{\mathfrak{h}} |\eta(H)| \, dH \int_{G/Z} \hat{f}(y.H) \cdot g(xy.H) \, dy^*.$$

Hence

$$\int_{G/Z} dx^* \left| \int_{\mathfrak{g}} f(X) \cdot \hat{g}(x.X) \, dX \right| \leq$$

$$\int_{\mathfrak{h}} |\eta(H)| \, dH \int_{G/Z \times G/Z} \left| \hat{f}(y.H) \right| \cdot \left| g(x.H) \right| dx^* \, dy^*.$$

Put $\alpha = |\hat{f}|$ and $\beta = |g|$. Then α and β lie in $\mathcal{D}$, and they are positive. Note that for $H \in \mathfrak{h}'$

$$|\eta(H)| \int_{G/Z} dx^* \int_{G/Z} \left| \hat{f}(y.H) \cdot g(xy.H) \right| dy^* = \phi_\alpha(H) \cdot \phi_\beta(H).$$

It is clear that $\phi_\beta \in C_c^\infty(\mathfrak{h}')$, and therefore

$$\int_{G/Z} dx^* \left| \int_{\mathfrak{g}} f(X) \cdot \hat{g}(x.X) \, dX \right| \leq \int_{\mathfrak{h}} \phi_\alpha(H) \cdot \phi_\beta(H) \, dH < \infty. \quad \square$$

COROLLARY 1.16. *Put $\omega = \mathrm{Supp}(\hat{g})$. Then, for $f \in \mathcal{D}$*

$$\Theta \colon f \mapsto \int_{G/Z} dx^* \int_{\mathfrak{g}} f(X) \cdot \hat{g}(x.X) \, dX$$

is a distribution in $J(\omega)$.

PROOF. This is obvious.   □

By a cusp form on $\mathfrak{g}$ we mean an element $f \in \mathcal{D}$ with the following property. If Γ is any Cartan subalgebra of $\mathfrak{g}$ which is not elliptic, then $\phi_f^\Gamma = 0$. We denote by $^0\mathcal{D}$ the space of all cusp forms.

LEMMA 1.17. *Suppose* $g \in C_c^\infty(\mathfrak{g}_\mathfrak{h})$, *then* $\hat{g} \in {}^0\mathcal{D}$.

PROOF. Let Γ be a non-elliptic Cartan subalgebra of $\mathfrak{g}$. Put $A = A_\Gamma$ and fix $P \in \mathcal{P}(A)$ with $P = MN$. Then $(\mathfrak{m} + \mathfrak{n}) \cap \mathfrak{g}_\mathfrak{h} = \emptyset$ [**21**, Lemma 48, p. 108]. Hence

$$g_P = 0.$$

But then $\hat{g}_P = (g_P)\hat{} = 0$. Therefore

$$\phi_{\hat{g}}^\Gamma = \phi_{\hat{g}_P}^\Gamma = 0. \quad \square$$

THEOREM 1.18 (Theorem 9). *Define* Θ *as in Corollary 1.16.* Θ *is a locally summable function on* $\mathfrak{g}$ *which is locally constant on* $\mathfrak{g}'$.

This is proved in the same way as [**21**, Theorem 16, p. 92]. The fact that $\hat{g}$ is a cusp form plays a crucial role in this proof.

1.6. Completion of the proof of Theorem 1.1. As we have seen in §1.4, we may assume that $\mathfrak{h}$ is elliptic. Fix $H_0 \in \mathfrak{h}'$.

LEMMA 1.19 (Lemma 2). *Fix an open compact set* ω *in* $\mathfrak{g}$. *There exists* $g \in C_c^\infty(\mathfrak{g}_\mathfrak{h})$ *such that*

$$\Theta(f) = \phi_{\hat{f}}(H_0)$$

for all $f \in C_c^\infty(\omega)$. *Here* Θ *is defined as in Corollary 1.16.*

PROOF. By Corollary 1.3 we can choose an open neighborhood ω_0 of H_0 in $\mathfrak{h}'$ such that $\phi_{\hat{f}}$ is constant on ω_0 for every $f \in C_c^\infty(\omega)$.

The mapping $(x, H) \mapsto x.H$ from $G \times \omega_0$ into $\mathfrak{g}$ is everywhere submersive (since $\omega_0 \subset \mathfrak{h}'$). Hence ω_0^G is open in $\mathfrak{g}$ and from [**21**, Theorem 11, p. 49] we have a surjective, linear mapping

$$\alpha \mapsto f_\alpha$$

of $C_c^\infty(G \times \omega_0)$ onto $C_c^\infty(\omega_0^G)$ such that

$$\int_{G \times \omega_0} \alpha(x : H) \cdot F(x.H) \, dx \, dH = \int_\mathfrak{g} f_\alpha(X) \cdot F(X) \, dX$$

for any $F \in C_c(\mathfrak{g})$. Fix $\alpha \in C_c^\infty(G \times \omega_0)$ such that

$$\int_{G \times \omega_0} |\eta(H)|^{-1/2} \cdot \alpha(x : H) \, dx \, dH = 1.$$

For $H \in \mathfrak{h}'$, put

$$\beta(H) = |\eta(H)|^{-1/2} \int_G \alpha(x : H) \, dx.$$

Then $\beta \in C_c^\infty(\omega_0)$ and

$$\int_{\omega_0} \beta(H)\,dH = 1.$$

Now put $g = f_\alpha$. Then $g \in C_c^\infty(\omega_0^G) \subset C_c^\infty(\mathfrak{g}_\mathfrak{h})$. Therefore Lemma 1.15 is applicable. Fix $f \in \mathcal{D}$. Then

$$\int_{G/Z} dx^* \int_\mathfrak{g} f(X) \cdot \hat{g}(x.X)\,dX = \int_{G/Z} dx^* \int_\mathfrak{g} \hat{f}(X) \cdot g(x.X)\,dX$$

$$= \int_{G/Z} dx^* \int_\mathfrak{g} \hat{f}(x^{-1}.X) \cdot g(X)\,dX,$$

and

$$\int_\mathfrak{g} \hat{f}(x^{-1}.X) \cdot g(X)\,dX = \int_{G \times \omega_0} \hat{f}(x^{-1}y.H) \cdot \alpha(y:H)\,dy\,dH$$

since $g = f_\alpha$. But

$$\int_{G/Z} dx^* \int_{G \times \omega_0} \left|\hat{f}(x^{-1}y.H)\right| \cdot \left|\alpha(y:H)\right|\,dy\,dH < \infty.$$

Therefore

$$\int_{G/Z} dx^* \int_\mathfrak{g} f(X) \cdot \hat{g}(x.X)\,dX = \int_{\omega_0} \phi_{\hat{f}}(H) \cdot \beta(H)\,dH.$$

Now suppose $f \in C_c^\infty(\omega)$. Then since $\beta \in C_c^\infty(\omega_0)$ and $\phi_{\hat{f}}$ is constant on ω_0,

$$\int_{\omega_0} \phi_{\hat{f}}(H) \cdot \beta(H)\,dH = \phi_{\hat{f}}(H_0) \int_{\omega_0} \beta(H)\,dH = \phi_{\hat{f}}(H_0).$$

This shows that, for $f \in C_c^\infty(\omega)$,

$$\phi_{\hat{f}}(H_0) = \int_{G/Z} dx^* \int_\mathfrak{g} f(X) \cdot \hat{g}(x.X)\,dX. \quad \square$$

The first statement of Theorem 1.1 is now a consequence of Theorem 1.18 and Lemma 1.19. Moreover, the second statement of Theorem 1.1 would follow from Lemma 1.19 as soon as we can verify that $|\eta|^{1/2} \cdot \Theta$ is bounded on $\mathfrak{g}$.

2. Some results about neighborhoods of semisimple elements

By a G-domain in $\mathfrak{g}$, we mean a G-invariant subset of $\mathfrak{g}$ which is both open and closed. Note that the union and intersection of a finite number of G-domains is again a G-domain. By an orbit (or more precisely a G-orbit) in $\mathfrak{g}$ we mean a set of the form X^G where X is an element of $\mathfrak{g}$. For a G-orbit $\mathcal{O}$ in $\mathfrak{g}$, put

$$d(\mathcal{O}) = \dim\big(\mathfrak{g}/C_\mathfrak{g}(X)\big) \quad \text{and} \quad r(\mathcal{O}) = \dim C_\mathfrak{g}(X)$$

where $C_\mathfrak{g}(X)$ is the centralizer in $\mathfrak{g}$ of a point $X \in \mathcal{O}$. We call $d(\mathcal{O})$ the dimension and $r(\mathcal{O})$ the rank of $\mathcal{O}$. Let $\mathcal{O}(0)$ denote the set of all nilpotent orbits in $\mathfrak{g}$. Let $\mathcal{N}$ be the set of all nilpotent elements in $\mathfrak{g}$.

As we have already discussed in the Introduction, for some integer n, we may regard G and $\mathfrak{g}$ as subsets of $M_n(\Omega)$. For $X \in M_n(\Omega)$, put

$$|X| = \max_{i,j} |X_{ij}|$$

where the X_{ij} are the coefficients of the matrix. This defines, in particular, a norm on $\mathfrak{g}$.

LEMMA 2.1. *Fix* $Z \in \mathfrak{z}$ *and a neighborhood* ω *of* Z *in* $\mathfrak{g}$. *Let* $\mathfrak{h}_1, \mathfrak{h}_2, \ldots, \mathfrak{h}_r$ *be a complete set of Cartan subalgebras of* $\mathfrak{g}$ *no two of which are conjugate under* G. *Then there exists a* G-*domain* D *in* $\mathfrak{g}$ *such that* $Z \in D \subset \omega^G$ *and* $D \cap \mathfrak{h}_i \subset \omega$ *for* $1 \le i \le r$.

PROOF. By translating the entire problem by $-Z$, we are reduced to the case $Z = 0$. We may assume that ω is open. Define the polynomial functions $p_\ell, p_{(\ell+1)}, \ldots, p_n$ on $\mathfrak{g}$ by

$$\det\big(t - \mathrm{ad}(X)\big) = \sum_{k=\ell}^{n} p_k(X) t^k$$

where $X \in \mathfrak{g}$ and t is an indeterminate. For $\varepsilon > 0$, let $\mathfrak{g}(\varepsilon)$ denote the set of all elements in $\mathfrak{g}$ of the form $Z + X$ where $Z \in \mathfrak{z}$ and $X \in \mathfrak{g}_1 = [\mathfrak{g}, \mathfrak{g}]$ such that $|Z| \le \varepsilon$ and $|p_k(X)| \le \varepsilon$ for $\ell \le k \le n$. $\mathfrak{g}(\varepsilon)$ is certainly a G-domain in $\mathfrak{g}$ (the p_k are G-invariant), and $0 \in \mathfrak{g}(\varepsilon)$. We may choose ε small enough so that $\mathfrak{g}(\varepsilon) \cap \mathfrak{h}_i \subset \omega$ for $1 \le i \le r$. Note that if $X \in \mathfrak{g}(\varepsilon)$ is semisimple, then there exists $h \in G$ such that $X^h \in \mathfrak{h}_i$ for some i. This implies that $X^h \in \mathfrak{g}(\varepsilon) \cap \mathfrak{h}_i$. Therefore $X \in \omega^G$.

Choose $X \in \mathfrak{g}(\varepsilon)$. Let $\mathcal{O} = X^G$. Then $\mathrm{cl}(\mathcal{O})$ contains a semisimple element, X_0. Since $\mathfrak{g}(\varepsilon)$ is closed, $X_0 \in \mathfrak{g}(\varepsilon)$. But since X_0 is semisimple, this implies that $X_0 \in \omega^G$. Therefore, since ω^G is open, $\mathcal{O} \cap \omega^G \ne \emptyset$. But then $\mathcal{O} \subset \omega^G$, and so $X \in \omega^G$. Therefore $\mathfrak{g}(\varepsilon) \subset \omega^G$. $\square$

COROLLARY 2.2. *Let* γ *be a semisimple element of* $\mathfrak{g}$. *Let* $M = C_G(\gamma)$ *and* $\mathfrak{m} = C_{\mathfrak{g}}(\gamma)$. *Fix a neighborhood* ω *of* γ *in* $\mathfrak{m}$ *and let* $\mathfrak{h}_1, \mathfrak{h}_2, \ldots, \mathfrak{h}_r$ *be a complete set of Cartan subalgebras of* $\mathfrak{m}$ *no two of which are conjugate under* M. *Then there exists an* M-*domain* U *in* $\mathfrak{m}$ *such that* $\gamma \in U \subset \omega^M$ *and* $U \cap \mathfrak{h}_i \subset \omega$ *for* $1 \le i \le r$.

PROOF. M is reductive with Lie algebra $\mathfrak{m}$. Since γ is in the center of $\mathfrak{m}$, we may immediately apply Lemma 2.1. $\square$

COROLLARY 2.3. *Let* γ *be a semisimple element of* $\mathfrak{g}$. *Let* $M = C_G(\gamma)$ *and* $\mathfrak{m} = C_{\mathfrak{g}}(\gamma)$. *Fix a neighborhood* ω *of* γ *in* $\mathfrak{m}$ *and let* $\mathfrak{h}_1, \mathfrak{h}_2, \ldots, \mathfrak{h}_r$ *be a complete set of Cartan subalgebras of* $\mathfrak{m}$ *no two of which are conjugate under* M. *Then there exists an* M-*domain* U *in* $\mathfrak{m}$ *such that* $\gamma \in U \subset \omega^M$ *and* $U \cap \mathfrak{h}_i \subset \omega$ *for* $1 \le i \le r$. *Furthermore,* U *may be chosen so that*

(1) $\det\big(\mathrm{ad}(Y)\big)_{\mathfrak{g}/\mathfrak{m}} \ne 0$ *for all* $Y \in U$ *and*

(2) *for all compact sets* Q *of* $\mathfrak{g}$, *there exists a compact set* C *in* G *such that:*

$$U^x \cap Q = \emptyset \text{ unless } x \in CM.$$

PROOF. For each Cartan subalgebra $\mathfrak{h}_i$ as in Corollary 2.2 we can choose a neighborhood ω_i of γ in ω such that

(A) $\det\big(\mathrm{ad}(Y)\big)_{\mathfrak{g}/\mathfrak{m}} \ne 0$ *for all* $Y \in \omega_i$ *and*

(B) given any compact set Q of $\mathfrak{g}$, there exists a compact set C_i in G such that

$$(\omega_i \cap \mathfrak{h}_i)^x \cap Q = \emptyset \text{ unless } x \in C_i M.$$

By making ω_i small enough, (A) can be satisfied. For (B), see [**21**, Lemma 25, p. 64].

We may replace ω by $\cap_{i=1}^r \omega_i$ in Corollary 2.2 and choose U accordingly. Condition (1) is automatically satisfied. In order to verify condition (2) we may assume Q is open. Consider the map

$$\phi \colon G \times U \longrightarrow \mathfrak{g}$$

where $(x, Y) \mapsto Y^x$. Since condition (1) holds for U, it is easy to show that ϕ is everywhere submersive.

Fix a compact open subgroup K of G.

Suppose $Y_0^{x_0} \in Q$ for some $Y_0 \in U$ and $x_0 \in G$. Since $Kx_0 \times U$ is open in $G \times U$, its image is open in $\mathfrak{g}$. Therefore

$$\phi(Kx_0 \times U) \cap Q \cap \mathfrak{g}' \neq \emptyset,$$

and so there exists $Y \in U$ and $x \in Kx_0$ such that $Y^x \in Q \cap \mathfrak{g}'$.

Therefore $\eta_{\mathfrak{g}}(Y) \neq 0$. Since $\eta_{\mathfrak{g}/\mathfrak{m}}(Y) \neq 0$, we have $\eta_{\mathfrak{m}}(Y) \neq 0$. Hence Y is regular in $\mathfrak{m}$ and so $Y = H^m$ for some H and i such that

$$H \in U \cap \mathfrak{h}_i' \subset \omega \cap \mathfrak{h}_i.$$

Therefore

$$Y^x = H^{xm} \in (\omega \cap \mathfrak{h}_i)^{xm} \cap Q.$$

By condition (B) there exists a compact set $C = KC$ of G such that

$$(\omega \cap \mathfrak{h}_i)^y \cap Q = \emptyset$$

for $1 \leq i \leq r$ unless $y \in CM$. Therefore $xm \in CM$ and so $x_0 \in Kx \subset KCM = CM$. □

COROLLARY 2.4. *Suppose U satisfies the conditions of Corollary 2.2 and Corollary 2.3. Then $V = U^G$ is a G-domain in $\mathfrak{g}$.*

PROOF. The map $\phi \colon G \times U \longrightarrow \mathfrak{g}$ is submersive with image V, hence V is open. We need to verify that V is closed. Suppose $\{u_i\}$ and $\{x_i\}$ are sequences in U and G such that $u_i^{x_i} \to X \in \mathfrak{g}$. By condition (2) of Corollary 2.3, we can choose a compact set C in G such that $x_i \in CM$. Then $x_i = y_i m_i$ with $y_i \in C$ and $m_i \in M$. By selecting a subsequence, we may assume that $y_i \to y$. Then $u_i^{m_i} \to y^{-1}.X$. Since U is closed, $y^{-1}.X \in U$ and hence $X \in U^y \subset V$. This proves V is closed. □

REMARK 2.5. It is clear that $V \subset \omega^G$.

LEMMA 2.6. *Let γ be a semisimple element in $\mathfrak{g}$ and ω a neighborhood of γ in $\mathfrak{g}$. Then we can choose a G-domain V in $\mathfrak{g}$ such that $\gamma \in V \subset \omega^G$.*

PROOF. This is clear from the above. □

DEFINITION 2.7. Let $\mathcal{O}(\gamma)$ be the set of all G-orbits $\mathcal{O}$ in $\mathfrak{g}$ such that $\gamma \in \text{cl}(\mathcal{O})$.

LEMMA 2.8. The set $\mathcal{O}(\gamma)$ is exactly the set of all orbits $\mathcal{O}$ of the form $\mathcal{O} = (\gamma + Y)^G$ where Y is a nilpotent element of $\mathfrak{m}$.

PROOF. Define $\mathfrak{h}_1, \mathfrak{h}_2, \ldots, \mathfrak{h}_r$ as in Corollary 2.2. For any i, the set $\gamma^G \cap \mathfrak{h}_i$ is finite. Hence we can choose a neighborhood ω of γ in $\mathfrak{m}$ such that $\gamma^G \cap \mathfrak{h}_i \cap \omega = \{\gamma\}$ for $1 \leq i \leq r$. Define U so that the conditions of Corollary 2.2 and Corollary 2.3 are fulfilled with respect to this ω. Put $V = U^G$.

Let $\mathcal{O} \in \mathcal{O}(\gamma)$. Since V is open in $\mathfrak{g}$ and $\gamma \in V$, we have $\mathcal{O} \cap V \neq \emptyset$. Hence $\mathcal{O} \subset V$, and therefore $\mathcal{O} \cap U \neq \emptyset$. Choose $Y \in \mathfrak{m}$ such that $\gamma + Y \in \mathcal{O} \cap U$. Since $\gamma \in \text{cl}(\mathcal{O})$, we can find a sequence $\{x_j\}$ in G such that $(\gamma + Y)^{x_j} \to \gamma$. Since U satisfies condition (2) of Corollary 2.3, we can select a subsequence so that $x_j = y_j m_j$ where $m_j \in M$ and $y_j \to y$ in G. Then

$$(\gamma + Y)^{m_j} \to y^{-1}.\gamma.$$

But $(\gamma + Y)^{m_j} \in U$ which is closed. Hence $y^{-1}.\gamma$ is a semisimple element of U. Therefore we can choose $m \in M$ and an index i such that $my^{-1}.\gamma \in \mathfrak{h}_i$. Then

$$(\gamma + Y)^{mm_j} \to my^{-1}.\gamma \in \mathfrak{h}_i \cap U \subset \mathfrak{h}_i \cap \omega.$$

Since $\gamma^G \cap \mathfrak{h}_i \cap \omega = \{\gamma\}$, we conclude that $my^{-1}.\gamma = \gamma$. Therefore $\gamma \in \text{cl}(\gamma + Y)^M$, and consequently $0 \in \text{cl}(Y^M)$. This proves that Y is a nilpotent element of $\mathfrak{m}$.

Conversely, if Y is a nilpotent element of $\mathfrak{m}$, let $\mathcal{O} = (\gamma + Y)^G$. Since Y is nilpotent, there exists a sequence $\{m_j\}$ in M such that $Y^{m_j} \to 0$. Therefore $(\gamma + Y)^{m_j} \to \gamma$ and so $\gamma \in \text{cl}(\mathcal{O})$. $\square$

LEMMA 2.9. Define U as above. Suppose $\mathcal{O} \in \mathcal{O}(\gamma)$. Then $\mathcal{O} \cap U = \gamma + \xi$ where ξ is a nilpotent M-orbit in $\mathfrak{m}$.

PROOF. Suppose $\mathcal{O} \in \mathcal{O}(\gamma)$. Then there exists a nilpotent Z in $\mathfrak{m}$ such that $\mathcal{O} = (\gamma + Z)^G$. Suppose $(\gamma + Z)^x \in \mathcal{O} \cap U$. As in the previous lemma, there exists a nilpotent Y in $\mathfrak{m}$ such that $(\gamma + Z)^x = \gamma + Y$. Since the Jordan decomposition is unique, $\gamma^x = \gamma$. This implies $x \in M$ and so $\mathcal{O} \cap U = (\gamma + Y)^M = \gamma + \xi$. $\square$

COROLLARY 2.10. There is a one-to-one correspondence between nilpotent M-orbits ξ in $\mathfrak{m}$ and elements $\mathcal{O}$ of $\mathcal{O}(\gamma)$ given by

$$\mathcal{O} = (\gamma + \xi)^G, \quad \mathcal{O} \cap U = \gamma + \xi.$$

PROOF. Clear. $\square$

COROLLARY 2.11. The set $\mathcal{O}(\gamma)$ has finite cardinality.

PROOF. Since $\mathcal{O}_\mathfrak{m}(0)$ is finite, this follows immediately from Corollary 2.10. $\square$

COROLLARY 2.12. We have

$$r(\mathcal{O}) = r(\xi).$$

PROOF. Suppose $\xi = Y^M$. Put $X = \gamma + Y$. Then $C_G(X) = C_M(Y)$. Therefore $r(\mathcal{O}) = \dim C_G(X) = \dim C_M(Y) = r(\xi)$.   □

LEMMA 2.13. *Suppose $\mathcal{O}_i \in \mathcal{O}(\gamma)$ and $\mathcal{O}_i \cap U = \gamma + \xi_i$. Then $\xi_1 \subset \mathrm{cl}(\xi_2)$ if and only if $\mathcal{O}_1 \subset \mathrm{cl}(\mathcal{O}_2)$.*

PROOF. Fix $Y_i \in U$ such that $\xi_i = Y_i^M$. If $Y_1 \in \mathrm{cl}(Y_2^M)$, it is clear that $\mathcal{O}_1 \subset \mathrm{cl}(\mathcal{O}_2)$. Now suppose $\mathcal{O}_1 \subset \mathrm{cl}(\mathcal{O}_2)$. Then we can choose a sequence $\{x_j\}$ in G such that $(\gamma + Y_2)^{x_j} \to (\gamma + Y_1)$. Because of the conditions on U, there exists a compact set C in G such that $x_j = y_j m_j$ with $y_j \in C$ and $m_j \in M$. By choosing a subsequence we have $y_j \to y$ in G and hence $(\gamma + Y_2)^{m_j} \to (\gamma + Y_1)^{y^{-1}}$. But then $(\gamma + Y_1)^{y^{-1}} \in U$, and so, as before, $y \in M$. Hence $m_j' = y m_j \in M$, and so $\xi_1 \subset \mathrm{cl}(\xi_2)$.   □

Put $W = \cup_{\mathcal{O} \in \mathcal{O}(\gamma)} \mathcal{O}$, and, for any integer d, let W_d denote the union of all orbits $\mathcal{O} \in \mathcal{O}(\gamma)$ with $d(\mathcal{O}) \leq d$.

LEMMA 2.14. *The set W_d is closed in $\mathfrak{g}$. Moreover, if $\mathcal{O} \in \mathcal{O}(\gamma)$ and $d(\mathcal{O}) = d$, then $\mathrm{cl}(\mathcal{O}) \subset \mathcal{O} \cup W_{d-1}$.*

PROOF. This is obvious from the above lemma and general facts about nilpotent orbits of reductive groups.   □

3. Proof of Theorem 3.1

Let $\mathcal{D}_0$ be the space of all $f \in \mathcal{D}$ such that $\phi_f^{\mathfrak{h}} = 0$ for all Cartan subalgebras $\mathfrak{h}$ of $\mathfrak{g}$. The next step is to prove the following Theorem.

THEOREM 3.1 (Theorem 10). *We have $T(f) = 0$ for $T \in J$ and $f \in \mathcal{D}_0$.*

3.1. A homogeneity result. For $f \in \mathcal{D}$, $X \in \mathfrak{g}$, and $t \in \Omega^\times$ define $f_t(X) = f(t^{-1}X)$. If T is a distribution on $\mathfrak{g}$, define the distribution $\rho(t)T$ by[2]

$$\langle \rho(t)T, f \rangle = \langle T, f_t \rangle$$

for $f \in \mathcal{D}$.

Fix an orbit $\mathcal{O}$ and a point $X_0 \in \mathcal{O}$. Let $C_G(X_0)$ denote the centralizer of X_0 in G. Then $C_G(X_0)$ is unimodular and therefore the homogeneous space $G/C_G(X_0)$ has an invariant measure dx^* which is unique up to a constant factor. By a theorem of Deligne and Rao [**55**] the integral

$$\int_{G/C_G(X_0)} f(xX) \, dx^*$$

is well defined for $f \in \mathcal{D}$. Let $\mu_{\mathcal{O}}(f)$ denote its value. Then $\mu_{\mathcal{O}}$ is a positive measure on $\mathfrak{g}$ which is uniquely determined by the orbit $\mathcal{O}$ up to a constant factor.

Since $t\mathcal{N} = \mathcal{N}$, it is obvious that $J(\mathcal{N})$ is stable under $\rho(t)$. Moreover,

$$\mu_{\mathcal{O}}(f_t) = \int_{G/C_G(X_0)} f(t^{-1}x.X_0) \, dx^* = c\mu_{t^{-1}\mathcal{O}}(f).$$

―――――――――

[2]If T is a distribution on $\mathfrak{g}$ and $f \in \mathcal{D}$, it is sometimes convenient to write $\langle T, f \rangle$ for $T(f)$.

This shows that $\rho(t)\mu_{\mathcal{O}} = c\mu_{t^{-1}\mathcal{O}}$ (where $c = c(t) \in \mathbb{R}_+^\times$).

LEMMA 3.2 (Lemma 3). *Fix $\mathcal{O} \in \mathcal{O}(0)$. Then $\rho(t^2)\mu_{\mathcal{O}} = |t|^{d(\mathcal{O})}\mu_{\mathcal{O}}$.*

PROOF. If $d = d(\mathcal{O}) = 0$, then $\mathcal{O} = \{0\}$, and the statement is obvious. So we may assume $d > 0$. Fix $X_0 \in \mathcal{O}$ and complete it to (H_0, X_0, Y_0) by Jacobson-Morosow as usual. Let $f \in C_c(G)$. Then

$$\int_G f(x)\,dx = \int_{G/C_G(X_0)} dx^* \int_{C_G(X_0)} f(xz)\,dz.$$

Let $\mathbf{L}$ be the connected algebraic subgroup of $\mathbf{G}$ corresponding to $\bar{\Omega}H_0 + \bar{\Omega}X_0 + \bar{\Omega}Y_0$. Put $L = \mathbf{L} \cap G$ ($\mathbf{L}$ is defined over Ω). Let Γ be the Cartan subgroup of L corresponding to ΩH_0. Since H_0 normalizes X_0, Γ normalizes $C_G(X_0)$ and so acts on $G^* = G/C_G(X_0)$ on the right. Put

$$g(x^*) = \int_{C_G(X_0)} f(xz)\,dz.$$

Then, if $\gamma \in \Gamma$,

$$g(x^*\gamma) = \int_{C_G(X_0)} f(x\gamma z)\,dz.$$

On the other hand

$$\int_G f(x)\,dx = \int_G f(x\gamma)\,dx = \int_{G^*} dx^* \int_{C_G(X_0)} f(xz\gamma)\,dz.$$

But

$$\int_{C_G(X_0)} f(xz\gamma)\,dz = \int_{C_G(X_0)} f(x\gamma z) \frac{dz^\gamma}{dz}\,dz$$

$$= \left|\det\big(\mathrm{Ad}(\gamma)\big)_{C_{\mathfrak{g}}(X_0)}\right| \int_{C_G(X_0)} f(x\gamma z)\,dz$$

$$= \left|\det\big(\mathrm{Ad}(\gamma)\big)_{C_{\mathfrak{g}}(X_0)}\right| \cdot g(x^*\gamma).$$

This shows that

$$\int_{G^*} g(x^*)\,dx^* = \left|\det\big(\mathrm{Ad}(\gamma)\big)_{C_{\mathfrak{g}}(X_0)}\right| \int_{G^*} g(x^*\gamma)\,dx^*$$

for γ in Γ. But

$$\det\big(\mathrm{Ad}(\gamma)\big)_{C_{\mathfrak{g}}(X_0)} = \prod_{i=1}^r \xi_i(\gamma)$$

in the notation of [21, Lemma 34, p. 74]. Hence, still using the notation of [21, Lemma 34],

$$\left|\det\big(\mathrm{Ad}(\gamma)\big)_{C_{\mathfrak{g}}(X_0)}\right| = \prod_{i=1}^r |\xi(\gamma)|^{\lambda_i/2}.$$

Now

$$\lambda_1 + \lambda_2 + \cdots + \lambda_r = \sum_{i=1}^r (\lambda_i + 1) - r = n - r = d.$$

Hence

$$\int_{G^*} g(x^*) \, dx^* = |\xi(\gamma)|^{d/2} \int_{G^*} g(x^*\gamma) \, dx^*$$

for $g \in C_c(G^*)$ and $\gamma \in \Gamma$. Now let $f \in \mathcal{D}$. Then

$$\mu_{\mathcal{O}}(f) = \int_{G^*} f(x^* X) \, dx^*.$$

Fix $\gamma \in \Gamma$ and put $t = \xi(\gamma)^{-1}$. Then $\gamma.X_0 = \xi(\gamma)X_0 = t^{-1}X_0$. Hence

$$\mu_{\mathcal{O}}(f_t) = \int_{G^*} f(x^*\gamma.X_0) \, dx^* = |\xi(\gamma)|^{-d/2} \int_{G^*} f(x^*.X_0) \, dx^*$$
$$= |t|^{d/2} \mu_{\mathcal{O}}(f).$$

By considering the adjoint representation of L we see that every element in $(\Omega^\times)^2$ can be written as $\xi(\gamma)$ for some $\gamma \in \Gamma$. This proves the lemma. $\square$

3.2. A basis for $J(\mathcal{N})$. For any integer d, let $\mathcal{N}_d$ denote the union of all nilpotent orbits $\mathcal{O}$ with $d(\mathcal{O}) \leq d$. If $\mathcal{O}$ is a nilpotent orbit of dimension d, then $\mathcal{O} \cup \mathcal{N}_{d-1}$ is closed and $\mathcal{O}$ is open in $\mathcal{N}_d$. Furthermore $\mathcal{N}_d$ is closed in $\mathfrak{g}$ and $\mathcal{N} = \emptyset$ for $d < 0$.

LEMMA 3.3 (Lemma 4). *The measures $\mu_{\mathcal{O}}$, indexed by $\mathcal{O} \in \mathcal{O}(0)$, form a base for $J(\mathcal{N})$.*

We need some preparation for the proof.

LEMMA 3.4. *Let X be a totally disconnected space and Y a closed subspace of X. For any $f \in C_c^\infty(X)$, let $\bar{f}$ denote its restriction to Y. Then $f \mapsto \bar{f}$ is a surjective linear mapping of $C_c^\infty(X)$ onto $C_c^\infty(Y)$.*

PROOF. Choose $g \in C_c^\infty(Y)$ and let $\omega = \mathrm{Supp}(g)$. For any $y \in \omega$, we can choose an open compact neighborhood U_y of y in X such that g is constant on $Y \cap U_y$. Since ω is compact, we can choose a finite number of elements $y_1, y_2, \ldots, y_r \in \omega$ such that the $U_i = U_{y_i}$ cover ω. Let

$$c_i = g(y_i) \text{ and } V_i = U_i - \bigcup_{1 \leq j < i} U_j.$$

Then the V_i are disjoint open and compact subsets of X and $\omega \subset \cup_{i=1}^r V_i$. Let

$$I = \{i \colon V_i \cap Y \neq \emptyset\}.$$

Define $f = \sum_{i \in I} c_i f_i$ where f_i is the characteristic function of V_i. Then $f \in C_c^\infty(X)$. We claim that $\bar{f} = g$. Choose $y \in Y$. If $y \notin \cup_{i \in I} V_i$, then $f(y) = g(y) = 0$. So now suppose $y \in V_i$ for some $i \in I$. Then $f(y) = c_i$. But

$$y \in V_i \cap Y \subset U_i \cap Y$$

which implies that $g(y) = c_i$. $\square$

LEMMA 3.5. *Let G be a totally disconnected group and H a closed subgroup such that $G^* = G/H$ has an invariant measure dx^*. Let λ denote the usual representation of G on $C_c^\infty(G^*)$ so that*

$$\bigl(\lambda(y)f\bigr)(x^*) = f(y^{-1}x^*)$$

for $y \in G, f \in C_c^\infty(G^*)$, and $x^* \in G^*$. Let

$$\mathcal{D}_1 = \{f \in C_c^\infty(G^*)\colon \int_{G^*} f(x^*)\,dx^* = 0\}$$

and let $\mathcal{D}_2$ be the linear span of all elements of the form

$$\lambda(x)\phi - \phi$$

where $x \in G$ and $\phi \in C_c^\infty(G^*)$. Then $\mathcal{D}_1 = \mathcal{D}_2$.

PROOF. Clearly $\mathcal{D}_2 \subset \mathcal{D}_1$. So let τ be a linear functional on $C_c^\infty(G^*)/\mathcal{D}_2$. We can regard τ as a distribution on G^*.

Let ℓ denote the left regular map of G on $\mathcal{D}$. Then we have a natural, surjective, linear G-map

$$\alpha \mapsto f_\alpha$$

from $\mathcal{D}$ onto $C_c^\infty(G^*)$ where

$$f_\alpha(x^*) = \int_H \alpha(xh)\,d_l h$$

and $d_l h$ is a left Haar measure on H.

Define $T(\alpha) = \tau(f_\alpha)$. T is a distribution on G. Further

$$T\bigl(\ell(y)\alpha\bigr) = \tau\bigl(\lambda(y)f_\alpha\bigr) = \tau(f_\alpha) = T(\alpha)$$

for all $y \in G$. By [**21**, Lemma 17, p. 50] this implies that there exists a $c \in \mathbb{C}$ such that $T(\alpha) = c \int_G \alpha(x)\,d_l x$. This implies that $\tau(f_\alpha) = T(\alpha) = c \int_{G^*} f_\alpha(x^*)\,dx^*$. Since $\alpha \mapsto f_\alpha$ is a surjection, we conclude that for all $f \in C_c^\infty(G^*)$

$$\tau(f) = c \int_{G^*} f(x^*)\,dx^*.$$

Therefore $\mathcal{D}_1 \subset \ker(\tau)$. Since τ was an arbitrary linear functional on $C_c^\infty(G^*)/\mathcal{D}_2$, we conclude that $\mathcal{D}_1 \subset \mathcal{D}_2$.   □

LEMMA 3.6. *Let f be an element in $\mathcal{D}$ such that*
  (1) *$\mu_{\mathcal{O}}(f) = 0$ for all $\mathcal{O} \in \mathcal{O}(0)$ and*
  (2) *$\mathrm{Supp}(f) \cap \mathcal{N}_d = \emptyset$.*
Then $T(f) = 0$ for all $T \in J(\mathcal{N})$.

Note that $\mathcal{N}_d = \mathcal{N}$ for $d \geq n - \ell$ and so this lemma is obviously true for $d \geq n - \ell$. Assuming it is true for d, we will now prove it holds for $(d-1)$. We may assume $d \geq 1$ since condition (1) implies $f(0) = 0$ which implies condition (2) when $d \leq 0$.

Let $\mathcal{O}_1, \mathcal{O}_2, \ldots, \mathcal{O}_p$ be all the distinct nilpotent orbits of dimension d. Put

$$\mathcal{N}(i) = \mathcal{N}_{d-1} \cup \bigcup_{i < j \leq p} \mathcal{O}_j.$$

Then $\mathcal{N}(0) = \mathcal{N}_d$ and $\mathcal{N}(p) = \mathcal{N}_{d-1}$. Hence, this lemma will follow if we can prove the following lemma by induction on i.

LEMMA 3.7. *Fix an integer i with $0 \leq i \leq p$, and let f be an element in $\mathcal{D}$ such that*

(1) $\mu_{\mathcal{O}}(f) = 0$ *for all $\mathcal{O} \in \mathcal{O}(0)$ and*
(2) $\mathrm{Supp}(f) \cap \mathcal{N}(i) = \emptyset$.

Then $T(f) = 0$ for all $T \in J(\mathcal{N})$.

PROOF. We proceed by induction. We shall assume the case $(i-1)$ and prove the case i. We note that $\mathcal{N}(i-1) = \mathcal{N}(i) \cup \mathcal{O}_i$. Therefore if $\mathrm{Supp}(f) \cap \mathcal{O}_i = \emptyset$, it is clear that $\mathrm{Supp}(f) \cap \mathcal{N}(i-1) = \emptyset$ and our assertion follows by the induction hypothesis. Hence we may suppose that $\mathrm{Supp}(f) \cap \mathcal{O}_i \neq \emptyset$. Fix $X_0 \in \mathrm{Supp}(f) \cap \mathcal{O}$ where $\mathcal{O} = \mathcal{O}_i$. Since $d \geq 1$, $X_0 \neq 0$ and so by the Jacobson-Morosow lemma we can complete it to (H_0, X_0, Y_0). Now, $\mathrm{cl}(\mathcal{O}) \subset \mathcal{O} \cup \mathcal{N}_{d-1}$ and by hypothesis:

$$\mathrm{Supp}(f) \cap \mathcal{N}_{d-1} \subset \mathrm{Supp}(f) \cap \mathcal{N}(i) = \emptyset.$$

Therefore, $\mathrm{Supp}(f) \cap \mathrm{cl}(\mathcal{O}) = \mathrm{Supp}(f) \cap \mathcal{O}$ and so $\mathrm{Supp}(f) \cap \mathcal{O}$ is compact.

Let $U = C_{\mathfrak{g}}(Y_0)$. From [**21**, p. 76] the map:

$$\psi \colon G \times U \longrightarrow \mathfrak{g}$$

defined by $(x, u) \mapsto (X_0 + u)^x$ is submersive. Hence its image is an open G-invariant subset ω of X_0 in $\mathfrak{g}$. Further, $\mathcal{O}$ is open in $\mathcal{N}_d$. Therefore, we can find an open neighborhood ω_1 of X_0 in $\mathfrak{g}$ such that $\omega_1 \subset \omega$ and $\omega_1 \cap \mathcal{N}_d \subset \mathcal{O}$. Since $\psi \colon G \times \omega_1 \to \mathfrak{g}$ is submersive, we may assume that $\omega_1 = \omega_1^G$. Choose an open neighborhood U_0 of zero in U such that $X_0 + U_0 \subset \omega_1$ and $(X_0 + U_0) \cap X_0^G = X_0$ (see [**21**, Lemma 37]). Then $\omega_0 = (X_0 + U_0)^G \subset \omega_1^G = \omega_1$ is an open subset of ω_1 and $\omega_0 \cap \mathcal{N}_d = \mathcal{O}$. Further, $\mathcal{O}$ is closed in ω_0 since $\mathrm{cl}_{\mathfrak{g}}(\mathcal{O}) \subset \mathcal{N}_d$. Now

$$\mu_{\mathcal{O}}(f) = \int_{G/C_G(X_0)} f(x.X_0) \, dx^* = 0.$$

Hence, by Lemma 3.5, we can choose $\phi_1, \phi_2, \ldots, \phi_s \in C_c^\infty(\mathcal{O})$ and $y_1, y_2, \ldots, y_s \in G$ such that

$$f(x.X_0) = \sum_{j=1}^s \big(\phi_j(y_j x.X_0) - \phi_j(x.X_0)\big)$$

for $x \in G$. By Lemma 3.4 we can choose $\Phi_j \in C_c^\infty(\omega_0)$ such that $\phi_j = \Phi_j|_{\mathcal{O}}$. For $X \in \mathfrak{g}$, define

$$f_0(X) = \sum_{j=1}^s \big(\Phi_j(y_j.X) - \Phi_j(X)\big).$$

Then $f_0 \in C_c^\infty(\omega_0)$ and $f_0 = f$ on $\mathcal{O}$. Moreover, $T(f_0) = 0$ for any G-invariant distribution T on $\mathfrak{g}$. Finally, $\mathrm{Supp}(f_0) \cap \mathcal{N}_d \subset \omega_0 \cap \mathcal{N}_d = \mathcal{O}$. Put $f_1 = f - f_0$. Then $f_1 \in \mathcal{D}$ and $f_1 = 0$ on $\mathcal{O}$. Moreover

$$\mathrm{Supp}(f_1) \subset \mathrm{Supp}(f) \cup \mathrm{Supp}(f_0)$$

and since $\mathrm{Supp}(f) \cap \mathcal{N}(i) = \emptyset$, we conclude that

$$\mathrm{Supp}(f_1) \cap \mathcal{N}(i) \subset \omega_0 \cap \mathcal{N}(i) \subset \omega_0 \cap \mathcal{N}_d = \mathcal{O}.$$

But $f_1 = 0$ on $\mathcal{O}$, hence $\mathrm{Supp}(f_1) \cap \mathcal{N}(i-1) = \emptyset$.

Since $\mathrm{Supp}(f_1) \cap \mathcal{N}(i-1) = \emptyset$, f_1 satisfies the induction hypothesis and so $T(f_1) = 0$ for all $T \in J(\mathcal{N})$.

Hence $T(f) = T(f_0 + f_1) = T(f_0) + T(f_1) = 0$. $\qquad\square$

This lemma shows that the set $\{\mu_\mathcal{O} \colon \mathcal{O} \in \mathcal{O}(0)\}$ spans $J(\mathcal{N})$. We need only show that the elements of this set are linearly independent.

LEMMA 3.8. *The measures $\mu_\mathcal{O}$, indexed by $\mathcal{O} \in \mathcal{O}(0)$, are linearly independent.*

PROOF. Let $\mathcal{O}_1, \mathcal{O}_2, \ldots, \mathcal{O}_r$ be all the distinct elements of $\mathcal{O}(0)$ and let $d_i = d(\mathcal{O}_i)$. We assume that $d_1 \leq d_2 \leq \cdots \leq d_r$ so that, in particular, $\mathcal{O}_1 = \{0\}$ and $d_1 = 0$. Let $\mu_i = \mu_{\mathcal{O}_i}$ and suppose there exist $c_i \in \mathbb{C}$ such that

$$\sum_{i=1}^{r} c_i \mu_i = 0.$$

We need to show that $c_i = 0$ for $1 \leq i \leq r$. Suppose this is false. Let j be the highest index such that $c_j \neq 0$. Note that we may assume $j \geq 2$. Let $\mathcal{O} = \mathcal{O}_j$ and $d = d_j$. Fix $X_0 \in \mathcal{O}$. As before we can choose an open, G-invariant neighborhood ω_0 of X_0 in $\mathfrak{g}$ such that

$$\omega_0 \cap \mathcal{N}_d = \mathcal{O} \quad \text{and} \quad \mathrm{cl}_{\omega_0}(\mathcal{O}) = \mathcal{O}.$$

We can choose $f \in C_c^\infty(\omega_0)$ such that $f \geq 0$ and $f(X_0) = 1$. Then

$$\mu_j(f) = \mu_\mathcal{O}(f) > 0.$$

Now fix $i < j$. Then

$$\mathrm{Supp}(\mu_i) \subset \mathrm{cl}(\mathcal{O}_i) \subset \mathcal{O}_i \cup \mathcal{N}_{d_i - 1} \subset \mathcal{O}_i \cup \mathcal{N}_{d-1}.$$

Hence

$$\mathrm{Supp}(\mu_i) \subset \mathcal{N}_d \quad \text{and} \quad \mathrm{Supp}(\mu_i) \cap \mathcal{O} = \emptyset.$$

But

$$\mathrm{Supp}(\mu_i) \cap \omega_0 \subset \mathcal{N}_d \cap \omega_0 = \mathcal{O}.$$

Therefore $\mathrm{Supp}(\mu_i) \cap \omega_0 = \emptyset$ and so $\mu_i(f) = 0$. Hence

$$0 = \sum_{1 \leq i \leq j} c_i \mu_i(f) = c_j \mu_j(f).$$

But since $c_j \neq 0$ and $\mu_j(f) > 0$ we have a contradiction. $\qquad\square$

This completes the proof of Lemma 3.3.

3.3. Proof of Lemma 3.9. We shall prove Theorem 3.1 by induction on $\dim G$. Let $\mathfrak{z}$ denote the center of $\mathfrak{g}$. Fix a semisimple element γ in $\mathfrak{g}$ and let $\mathfrak{m}$ and M denote the centralizers of γ in $\mathfrak{g}$ and G, respectively. Assume $\gamma \notin \mathfrak{z}$. Then the induction hypothesis is applicable to M and one can prove the following lemma.

LEMMA 3.9 (Lemma 5). *There exists a G-domain V containing γ with the following property. If $\phi_f^{\mathfrak{h}} = 0$ for all $\mathfrak{h}$ in $\mathfrak{m}$ and $\mathrm{Supp}(f) \subset V$, then $T(f) = 0$ for all $T \in J$.*

PROOF. Let $\mathfrak{h}_1, \mathfrak{h}_2, \ldots, \mathfrak{h}_r$ be a complete set of Cartan subalgebras of $\mathfrak{m}$, no two of which are conjugate under M. Let W_{ji} denote the set of linear bijections $s \colon \mathfrak{h}_i \to \mathfrak{h}_j$ for which there exists an $x \in G$ with the property that $s(H) = x.H$ for all $H \in \mathfrak{h}_i$. As usual, x is called a representative of s in G. W_{ji} is a finite set.

If $s \in W_{ji}$ and $s(\gamma) = \gamma$, then the representative of s is in M and $s(\mathfrak{h}_i) = \mathfrak{h}_j$. In view of our assumption, this is impossible unless $i = j$. Note that $W_i = W_{ii}$ is a group. Let $W_i(M)$ be the subgroup of $s \in W_i$ such that $s(\gamma) = \gamma$.

For each $s \in W_{ji}$ choose a representative $x_s \in G$. Then $x_s \in M$ if and only if $s \in W_i(M)$ for some i.

Fix a compact open neighborhood ω of γ in $\mathfrak{g}$ such that $\omega \cap x_s(\omega) = \emptyset$ for $s \in W_{ji}$ unless $s(\gamma) = \gamma$.

Then, for $s \in W_{ji}$, unless $j = i$ and $s \in W_i(M)$, we have,

$$s(\omega \cap \mathfrak{h}_i) \cap \omega = x_s(\omega) \cap \omega \cap \mathfrak{h}_j = \emptyset \cap \mathfrak{h}_j = \emptyset.$$

Choose an M-domain U in $\mathfrak{m}$ such that the conditions of Corollary 2.2 and Corollary 2.3 are fulfilled with respect to $\omega \cap \mathfrak{m}$ (in place of ω). Put $V = U^G$. By Corollary 2.4, V is a G-domain.

The mapping $\phi \colon (x, Y) \mapsto Y^x$ of $G \times U$ into V is surjective and everywhere submersive. So by [**21**, Theorem 11, p. 49] we get a surjective linear mapping

$$\alpha \longmapsto f_\alpha$$

of $C_c^\infty(G \times U)$ onto $C_c^\infty(V)$ with the property that for all $F \in C(\mathfrak{g})$

$$\int_{G \times U} \alpha(x : Y) \cdot F(x.Y) \, dx \, dY = \int_V f_\alpha(X) \cdot F(X) \, dX.$$

Fix a G-invariant distribution T on V. Then by arguing as in [**21**, Lemma 21, p. 56] one can show that there exists an M-invariant distribution τ on U such that

$$\tau(\beta_\alpha) = T(f_\alpha)$$

for all $\alpha \in C_c^\infty(G \times U)$ where $\beta_\alpha \in C_c^\infty(U)$ is given by

$$\beta_\alpha(Y) = \int_G \alpha(x : Y) \, dx.$$

Now choose α such that $f_\alpha = f$ and put $\beta = \beta_\alpha$. Then for a G-invariant function h in $C(V)$ we have

$$\int_U \beta(Y) \cdot h(Y) \, dY = \int_{G \times U} \alpha(x : Y) \cdot h(Y) \, dY \, dx$$
$$= \int_{\mathfrak{g}} f(X) \cdot h(X) \, dX.$$

Since $V = U^G$, it is clear that

$$V \cap \mathfrak{g}' \subset \bigcup_{i=1}^r \mathfrak{g}_{\mathfrak{h}_i}.$$

Therefore

$$0 = \sum_{i=1}^r c_i \int_{\mathfrak{h}_i} |\eta(H)|^{1/2} \cdot h(H) \cdot \phi_f^{\mathfrak{h}_i}(H) \, d_i H$$
$$= \sum_{i=1}^r c_i \int_{\mathfrak{h}_i} |\eta(H)| \cdot h(H) \int_{G/A_i} f(x.H) \, d_i x^* \, d_i H$$
$$= \int_{\mathfrak{g}} f(X) \cdot h(X) \, dX.$$

This shows that

$$0 = \int_{\mathfrak{m}} \beta(Y) \cdot h(Y) \, dY = \sum_{i=1}^r c_i \int_{\mathfrak{h}_i} \phi_\beta^{\mathfrak{h}_i}(H) \cdot |\eta_{\mathfrak{m}}(H)|^{1/2} \cdot h(H) \, d_i H$$

where the c_i are certain positive constants and $\phi_\beta^{\mathfrak{h}_i} = \phi_\beta^{M/\mathfrak{h}_i}$.

Now fix i (say $i = 1$), and suppose $1 \le j \le r_0$ are all those indices j such that $\mathfrak{h}_j$ is conjugate to $\mathfrak{h} = \mathfrak{h}_1$ under G. Let $W = W_1$. Fix $h_0 \in C_c^\infty(\mathfrak{h}')$. There exists a G-invariant function $h \in C^\infty(\mathfrak{g})$ such that

(1) $h = 0$ outside $\mathfrak{g}_\mathfrak{h}$ and
(2) $h(xH) = \sum_{s \in W} h_0(s(H))$ for $x \in G$ and $H \in \mathfrak{h}$.

We then have

$$0 = \sum_{j=1}^{r_0} c_j \int_{\mathfrak{h}_j} \phi_\beta^{\mathfrak{h}_j}(H) \cdot |\eta_{\mathfrak{m}}(H)|^{1/2} \cdot h_0(H) \, d_j H$$

since $h = 0$ on $\mathfrak{g}_{\mathfrak{h}_i}$ for $i > r_0$. So

$$0 = \sum_{j=1}^{r_0} c_j \sum_{s \in W_{j1}} \int_{\mathfrak{h}} \phi_\beta^{\mathfrak{h}_j}(s(H)) \cdot |\eta_{\mathfrak{m}}(s(H))|^{1/2} \cdot h(H) \, dH.$$

Here we make use of the fact that $d(s_0(H)) = dH$ for $s_0 \in W$ since W is a finite group. Since this is true for all $h_0 \in C_c^\infty(\mathfrak{h}')$, we conclude that for $H \in \mathfrak{h}'$

$$0 = \sum_{j=1}^{r_0} c_j \sum_{s \in W_{j1}} \phi_\beta^{\mathfrak{h}_j}(s(H)) \cdot |\eta_{\mathfrak{m}}(s(H))|^{1/2}.$$

Now, $\text{Supp}(\phi_\beta^{\mathfrak{h}_j}) \subset U \cap \mathfrak{h}_j \subset \omega \cap \mathfrak{h}_j$. In particular, $\text{Supp}(\phi_\beta^{\mathfrak{h}}) \subset \omega \cap \mathfrak{h}$. Choose $H \in \omega \cap \mathfrak{h}'$. Then since $s(\omega \cap \mathfrak{h}_j) \cap \omega = \emptyset$ for $s \in W_{j1}$ unless $s \in W_{jj}(M)$ and $j = 1$, we have $\phi_\beta^{\mathfrak{h}_j}(sH) = 0$ unless $j = 1$ and $s \in W(M)$. Hence

$$0 = \sum_{s \in W(M)} \phi_\beta^{\mathfrak{h}}\big(s(H)\big) \cdot \big|\eta_{\mathfrak{m}}\big(s(H)\big)\big|^{1/2}$$

for $H \in \omega \cap \mathfrak{h}'$. But this shows that $\phi_\beta^{\mathfrak{h}}(H) = 0$ for $H \in \omega \cap \mathfrak{h}'$. This proves that $\phi_\beta^{\mathfrak{h}} = 0$. Since $\mathfrak{h}$ was chosen arbitrarily among $\mathfrak{h}_1, \mathfrak{h}_2, \ldots, \mathfrak{h}_r$, we conclude that for $1 \leq i \leq r$, we have $\phi_\beta^{\mathfrak{h}_i} = 0$.

We extend τ to a distribution on $\mathfrak{m}$ by defining

$$\tau(g) = \tau(F_U \cdot g)$$

for $g \in C_c^\infty(\mathfrak{m})$ where F_U is the characteristic function of U. Now applying the induction hypothesis of Theorem 3.1 to $\mathfrak{m}$, we conclude that $\tau(\beta) = 0$. Hence

$$T(f) = \tau(\beta) = 0. \quad \square$$

Recall that if $\mathcal{O}$ is any G-orbit, then $\text{cl}(\mathcal{O})$ contains a semisimple element. Moreover, the set of all G-domains is stable under complementation, finite unions, and finite intersections. Therefore the following corollary to Lemma 3.9 is immediate.

COROLLARY 3.10. *Let f be an element in $\mathcal{D}_0$ such that $\text{Supp}(f)$ does not meet $\mathfrak{z} + \mathcal{N}$. Then $T(f) = 0$ for all $T \in J$.*

PROOF. Let $\omega = \text{Supp}(f)$. Since $\omega \cap (\mathfrak{z} + \mathcal{N}) = \emptyset$, we can find for each γ in ω a G-domain V_γ such that

(1) $\gamma \in V_\gamma$ and
(2) for all $f \in C_c^\infty(V_\gamma)$ such that $\phi_f^{\mathfrak{h}} = 0$ for all Cartan subalgebras $\mathfrak{h}$ of $\mathfrak{g}$, $T(f) = 0$ for all $T \in J$,

by Lemma 3.9. Since ω is compact, there exists a finite number of elements $\gamma_1, \gamma_2, \ldots, \gamma_s$ in ω such that the $V_i = V_{\gamma_i}$ cover ω.

Let F_i denote the characteristic function of V_i. Let S denote the set $\{1, 2, \ldots, s\}$. For every nonempty subset I of S define

$$F_I = \prod_{i \in I} F_i \quad \text{and} \quad V_I = \bigcap_{i \in I} V_i.$$

Then $\prod_{1 \leq i \leq s}(1 - F_i) = 0$ on ω and so if $f_I = F_I \cdot f$, we have [3]

$$f = -\sum_{\emptyset \neq I \subset S} (-1)^{[I]} f_I.$$

But $\phi_{f_I}^{\mathfrak{h}}(H) = F_I(H) \cdot \phi_f^{\mathfrak{h}}(H) = 0$ for $H \in \mathfrak{h}'$. Since $f_I \in C_c^\infty(V_I) \subset C_c^\infty(V_i)$ for all $i \in I$, we have $T(f_I) = 0$ for all nonempty I in S. Hence $T(f) = 0$. $\qquad \square$

---

[3] $[S]$ denotes the number of elements in a set S.

3.4. Reduction to the semisimple case.

LEMMA 3.11. *In the proof of Theorem 3.1 we may assume that G is semisimple.*

PROOF. Since G is reductive, $\mathfrak{g} = \mathfrak{z} + \mathfrak{g}_1$ where $\mathfrak{g}_1 = [\mathfrak{g}, \mathfrak{g}]$ is a semisimple Lie algebra. Let $\mathbf{G_1}$ be the corresponding derived group of $\mathbf{G}$ and $\mathbf{Z}$ the connected algebraic subgroup of $\mathbf{G}$ corresponding to $\mathfrak{z}$. Let $G_1 = G \cap \mathbf{G_1}$ and $Z = \mathbf{Z} \cap G$. Then $G_0 = G_1 Z$ has finite index in G.

If $\dim \mathfrak{z} = 0$, we are done. Consequently, we may assume that $\dim \mathfrak{z} > 0$. Then, by induction, we may assume that Theorem 3.1 holds for G_1. Let Y be a complete set of representatives in G for G/G_0.

As usual, $\mathcal{D} = C_c^\infty(\mathfrak{z}) \otimes C_c^\infty(\mathfrak{g}_1)$. Suppose that $f \in \mathcal{D}$ and $\phi_f^{\mathfrak{h}} = 0$ for all Cartan subalgebras $\mathfrak{h}$ of $\mathfrak{g}$. We must show that $T(f) = 0$ for any G-invariant distribution T on $\mathfrak{g}$.

Let $f = \sum_{i=1}^r \alpha_i \otimes \beta_i$ where $\alpha_i \in C_c^\infty(\mathfrak{z})$ and $\beta_i \in C_c^\infty(\mathfrak{g}_1)$. Further, assume that $\alpha_1, \alpha_2, \ldots, \alpha_r$ are linearly independent over $\mathbb{C}$.

Define f_0 by

$$f_0 = \frac{1}{[Y]} \sum_{y \in Y} f^y = \sum_{i=1}^r \alpha_i \otimes \beta_i^0$$

where

$$\beta_i^0 = \frac{1}{[Y]} \sum_{y \in Y} \beta_i^y.$$

Note that $T(f_0) = T(f)$ and so it is enough to show that $T(f_0) = 0$.

Now

$$\phi_{f_0}^{\mathfrak{h}} = \phi_f^{\mathfrak{h}} = 0.$$

Hence if $H = Z + H_1$, then

$$0 = \phi_f^{\mathfrak{h}}(Z + H_1) = \sum_{i=1}^r \alpha_i(Z) \cdot \phi_{\beta_i^0}^{\mathfrak{h}_1}(H_1)$$

for $Z \in \mathfrak{z}$ and $H_1 \in \mathfrak{h}_1' = \mathfrak{h}' \cap \mathfrak{g}_1$. Since the α_i are linearly independent, it follows that $\phi_{\beta_i^0}^{\mathfrak{h}_1} = 0$ for $1 \leq i \leq r$.

Put $\tau_i(\beta) = T(\alpha_i \otimes \beta)$ for $\beta \in C_c^\infty(\mathfrak{g}_1)$. Then τ_i is a G_1-invariant distribution on $\mathfrak{g}_1$. Hence, by the induction hypothesis, $\tau_i(\beta_i^0) = 0$. Therefore

$$T(f) = T(f_0) = \sum_{i=1}^r \tau_i(\beta_i^0) = 0. \quad \square$$

So we now assume that G is semisimple and so $\mathfrak{z} = \{0\}$.

LEMMA 3.12 (Lemma 6). *Suppose $f \in \mathcal{D}_0$ such that $\mu_\mathcal{O}(f) = 0$ for all $\mathcal{O} \in \mathcal{O}(0)$. Then $T(f) = 0$ for $T \in J$.*

Lemma 3.12 will be proved if we can show that the following lemma holds for all d.

LEMMA 3.13. *Fix an integer d. Suppose $f \in \mathcal{D}$ such that*

(1) $\phi_f^{\mathfrak{h}} = 0$ *for all Cartan subalgebras* $\mathfrak{h}$ *of* $\mathfrak{g}$,
(2) $\mu_{\mathcal{O}}(f) = 0$ *for all* $\mathcal{O} \in \mathcal{O}(0)$, *and*
(3) $\mathrm{Supp}(f) \cap \mathcal{N}_d = \emptyset$.
Then $T(f) = 0$ *for all* $T \in J$.

When $d \geq n - \ell$, we have $\mathcal{N}_d = \mathcal{N}$ and so Corollary 3.10 makes this lemma true. If $d \leq 0$, then condition (2) implies condition (3). Hence we may assume that the lemma is true for d and show it is true for $d - 1$ where $1 \leq d \leq n - \ell$.

Again let $\mathcal{O}_1, \mathcal{O}_2, \dots, \mathcal{O}_p$ be all the distinct nilpotent orbits of dimension d. Let

$$\mathcal{N}(i) = \mathcal{N}_{d-1} \cup \bigcup_{i < j \leq p} \mathcal{O}_j$$

for $0 \leq i \leq p$. Since $\mathcal{N}(0) = \mathcal{N}_d$ and $\mathcal{N}(p) = \mathcal{N}_{d-1}$, this lemma will follow if we can prove the following lemma by induction on i.

LEMMA 3.14. *Fix an integer* i *with* $1 \leq i \leq p$. *Suppose* $f \in \mathcal{D}$ *such that*
(1) $\phi_f^{\mathfrak{h}} = 0$ *for all Cartan subalgebras* $\mathfrak{h}$ *of* $\mathfrak{g}$,
(2) $\mu_{\mathcal{O}}(f) = 0$ *for all* $\mathcal{O} \in \mathcal{O}(0)$, *and*
(3) $\mathrm{Supp}(f) \cap \mathcal{N}(i) = \emptyset$.
Then $T(f) = 0$ *for all* $T \in J$.

PROOF. We note that $\mathcal{N}(i-1) = \mathcal{N}(i) \cup \mathcal{O}_i$. Let $\mathcal{O} = \mathcal{O}_i$. If $\mathrm{Supp}(f) \cap \mathcal{O} = \emptyset$, then $\mathrm{Supp}(f) \cap \mathcal{N}(i-1) = \emptyset$, and our assertion follows by induction. Hence we may suppose that $\mathrm{Supp}(f) \cap \mathcal{O} \neq \emptyset$. Fix $X_0 \in \mathrm{Supp}(f) \cap \mathcal{O}$. Since $d \geq 1$, $X_0 \neq 0$.

As in the proof of Lemma 3.7, we may choose an open G-invariant set ω_0 such that
(1) $X_0 \in \omega_0$,
(2) $\omega_0 \cap \mathcal{N}_d = \mathcal{O}$, and
(3) $\mathcal{O}$ is closed in ω_0.
Furthermore, we can find f_0 and f_1 with the following properties.
(1) $f_0 + f_1 = f$.
(2) $f_0 \in C_c^\infty(\omega_0)$, $f_0 = f$ on $\mathcal{O}$, and $T(f_0) = 0$ for all $T \in J$.
(3) $f_1 \in \mathcal{D}$, $f_1|_{\mathcal{O}} = 0$, and $\mathrm{Supp}(f_1) \cap \mathcal{N}(i-1) = \emptyset$.
(4) $\phi_{f_1}^{\mathfrak{h}} = \phi_f^{\mathfrak{h}} = 0$ for all Cartan subalgebras $\mathfrak{h}$ in $\mathfrak{g}$.
Hence the induction hypothesis applies to f_1 and so $T(f_1) = 0$ for all $T \in J$. Therefore $T(f) = T(f_1) + T(f_0) = 0$ for all $T \in J$. $\square$

3.5. The proof of Theorem 3.1. Let

$$J_0 = \{T \in J : T(f) = 0 \text{ for all } f \in \mathcal{D}_0\}.$$

Recall that

$$\mathcal{D}_0 = \{f \in \mathcal{D} : \phi_f^{\mathfrak{h}} = 0 \text{ for all Cartan subalgebras } \mathfrak{h} \text{ of } \mathfrak{g}\}.$$

LEMMA 3.15 (Lemma 7). *We have* $J = J_0 + J(\mathcal{N})$.

PROOF. Put $J_1 = J_0 + J(\mathcal{N})$ and let $\mathcal{D}_1 \subset \mathcal{D}_0$ be the subspace of all $f \in \mathcal{D}_0$ such that $\mu_{\mathcal{O}}(f) = 0$ for all $\mathcal{O} \in \mathcal{O}(0)$. Then

$$\dim(\mathcal{D}_0/\mathcal{D}_1) \leq [\mathcal{O}(0)] < \infty.$$

Consider the bilinear form

$$(T, f) \longmapsto T(f)$$

on $J_1 \times \mathcal{D}_0$. Since $T(f) = 0$ for all $f \in \mathcal{D}_0$ implies $T \in J_0$ and similarly $T(f) = 0$ for all $T \in J_1$ implies $f \in \mathcal{D}_1$, we get a non-degenerate bilinear form on

$$J_1/J_0 \times \mathcal{D}_0/\mathcal{D}_1.$$

Let Λ be the space of all linear functionals on $\mathcal{D}_0/\mathcal{D}_1$. By the finiteness condition above,

$$\dim(\Lambda) = \dim(\mathcal{D}_0/\mathcal{D}_1) = \dim(J_1/J_0).$$

For any $T \in J$, let λ_T denote the restriction of T on $\mathcal{D}_0$. It follows from Lemma 3.12 that $\lambda_T \in \Lambda$. So, there exists a linear map $T \mapsto \lambda_T$ from J to Λ. The kernel of this map is J_0 and so

$$\dim(J/J_0) \leq \dim(\Lambda) = \dim(J_1/J_0) < \infty.$$

Consequently, $J_1 = J$. $\qquad\square$

LEMMA 3.16 (Lemma 8). *Suppose $f \in \mathcal{D}_0$ and $t \in \Omega^\times$. Then f_t and $\hat{f}$ are also in $\mathcal{D}_0$.*

PROOF. Let $\mathfrak{h}$ be a Cartan subalgebra of $\mathfrak{g}$. Fix $H_0 \in \mathfrak{h}'$. Then the distribution $T\colon f \mapsto \phi_{\hat{f}}^{\mathfrak{h}}(H_0)$ is a locally summable function on $\mathfrak{g}$. Hence

$$\phi_{\hat{f}}^{\mathfrak{h}}(H_0) = \int_{\mathfrak{g}} T(X) \cdot f(X)\, dX$$

$$= \sum_i c_i \int_{\mathfrak{h}_i} |\eta(H)|\, d_i H \int_{G/A_i} T(g.H) \cdot f(g.H)\, dg^*$$

$$= \sum_i c_i \int_{\mathfrak{h}_i} |\eta(H)|^{1/2} \cdot \phi_f^{\mathfrak{h}_i}(H) \cdot T(H)\, d_i H$$

in the usual notation. If $f \in \mathcal{D}_0$, then $\phi_f^{\mathfrak{h}_i} = 0$ and hence $\phi_{\hat{f}}^{\mathfrak{h}}(H_0) = 0$. Consequently, $\hat{f} \in \mathcal{D}_0$.

Also, for $H \in \mathfrak{h}'$,

$$0 = \phi_f^{\mathfrak{h}}(t^{-1}H) = |\eta(t^{-1}H)|^{1/2} \int_{G/A} f(x.(t^{-1}H))\, dx^*$$

$$= |t|^{\frac{\ell-n}{2}} \phi_{f_t}^{\mathfrak{h}}(H).$$

Consequently, $f \in \mathcal{D}_0$ implies $f_t \in \mathcal{D}_0$. $\qquad\square$

Now we come to the proof of Theorem 3.1.

In view of Lemma 3.15 it is enough to verify that $J(\mathcal{N}) \subset J_0$, or, from Lemma 3.3, that $\mu_{\mathcal{O}} \in J_0$ for all $\mathcal{O} \in \mathcal{O}(0)$.

Suppose this is false. Fix $\mathcal{O} \in \mathcal{O}(0)$ such that $T = \mu_{\mathcal{O}} \notin J_0$. Then by Lemma 3.16 we can choose $f \in \mathcal{D}_0$ such that $T(\hat{f}) \neq 0$. But since $\widehat{T}(f) = T(\hat{f})$, we conclude from Lemma 3.15 that $\widehat{T} = T_0 + S$ where $T_0 \in J_0$ and $S \in J(\mathcal{N})$. Hence, for $g \in \mathcal{D}_0$,

$$\widehat{T}(g) = S(g).$$

Fix $t \in \Omega^{\times}$. Then $f_t \in \mathcal{D}_0$ by Lemma 3.16. Hence

$$S(f_t) = \widehat{T}(f_t) = T\big((f_t)^{\widehat{}}\big).$$

But

$$\big((f_t)^{\widehat{}}\big)(X) = \int_{\mathfrak{g}} \chi\big(B(X,Y)\big) \cdot f(t^{-1}Y)\, dY$$

$$= |t|^n \int_{\mathfrak{g}} \chi\big(B(X,tY)\big) \cdot f(Y)\, dY = |t|^n\, \hat{f}(tX)$$

where $n = \dim(\mathfrak{g})$. Hence $(f_t)^{\widehat{}} = |t|^n\, (\hat{f})_{t^{-1}}$. Therefore

$$\big(\rho(t^2)S\big)(f) = S(f_{t^2}) = T\big((f_{t^2})^{\widehat{}}\big)$$

$$= |t|^{2n}\, T\big((\hat{f})_{t^{-2}}\big) = |t|^{2n-d}\, T(\hat{f})$$

$$= |t|^{2n-d}\, S(f)$$

from Lemma 3.2 where $d = d(\mathcal{O})$.

On the other hand,

$$S = \sum_i c_i\, \mu_i$$

where $c_i \in \mathbb{C}, \mu_i = \mu_{\mathcal{O}_i}$, and $\mathcal{O}_1, \mathcal{O}_2, \ldots, \mathcal{O}_s$ are the distinct nilpotent G-orbits. Therefore, by Lemma 3.2,

$$\rho(t^2)S = \sum_i c_i\, |t|^{d_i}\, \mu_i$$

where $d_i = d(\mathcal{O}_i)$. We conclude that

$$\sum_i c_i\, |t|^{d_i}\, \mu_i(f) = |t|^{2n-d} \sum_i c_i\, \mu_i(f).$$

Now we may assume that $\mathfrak{g} \neq \{0\}$. Then $\ell = \mathrm{rank}(\mathfrak{g}) > 0$. Put $r = r(\mathcal{O})$ and $r_i = r(\mathcal{O}_i)$. Then $r_i \geq \ell$ for every i and therefore

$$d_i - (2n - d) = -(r_i + r) \leq -2\ell < 0.$$

Since distinct quasi-characters of $\Omega^{\times}$ are linearly independent, we conclude from the above relation that

$$\sum_i c_i\, \mu_i(f) = 0.$$

But then

$$T(\hat{f}) = \widehat{T}(f) = S(f) = 0,$$

which contradicts the definition of f. This completes the proof of Theorem 3.1.

4. Some consequences of Theorem 3.1

We shall now derive some consequences of Theorem 3.1.

LEMMA 4.1 (Lemma 9). *Let V be a G-domain in $\mathfrak{g}$. Then the following three conditions on an element $f \in \mathcal{D}$ are equivalent.*

(1) $\mu_{\mathcal{O}}(f) = 0$ *for every regular orbit $\mathcal{O}$ in V.*
(2) $\mu_{\mathcal{O}}(f) = 0$ *for every orbit $\mathcal{O}$ in V.*
(3) $T(f) = 0$ *for all $T \in J(V)$.*

PROOF. Note that an orbit is called regular if it is contained in $\mathfrak{g}'$. Clearly (3) implies (2) implies (1).

Let F denote the characteristic function of V. Since V is both open and closed, $F \in C^{\infty}(\mathfrak{g})$. Define $f_V = F \cdot f$. Then $f_V \in \mathcal{D}$, and $V \cap \mathrm{Supp}(f - f_V) = \emptyset$. Hence $T(f) = T(f_V)$ for $T \in J(V)$.

From (1) we know that $\mu_{\mathcal{O}}(f_V) = 0$ for all regular orbits in $\mathfrak{g}$. But this means

$$\phi_{f_V}^{\mathfrak{h}}(H) = |\eta(H)|^{1/2} \int_{G/A} f_V(x.H)\, dx^* = 0$$

for all Cartan subalgebras $\mathfrak{h}$ of $\mathfrak{g}$ and $H \in \mathfrak{h}'$. Hence $f_V \in \mathcal{D}_0$, and, by Theorem 3.1, $T(f_V) = 0$ for all $T \in J$.

For $T \in J(V)$, $T(f) = T(f_V) = 0$. Therefore, condition (1) implies condition (3). □

COROLLARY 4.2. *Let ω be a compact set, L a lattice, and V a G-domain in $\mathfrak{g}$. Assume that $V \subset \mathrm{cl}(\omega^G)$. Then we can choose a finite number of regular orbits $\mathcal{O}_1, \mathcal{O}_2, \ldots, \mathcal{O}_r$ in V such that the $j_L(\mu_{\mathcal{O}_i})$ span $j_L J(V)$.*

PROOF. Since $V \subset \mathrm{cl}(\omega^G)$, we have $J(V) \subset J(\omega)$, and so, by Theorem 12.1, $\dim j_L J(V) < \infty$. Consequently, it is enough to show that if λ is a linear functional on $j_L J(V)$ such that $\lambda(\mu_{\mathcal{O},L}) = 0$ for every regular orbit in V, then $\lambda = 0$.

For any $f \in C_c(\mathfrak{g}/L)$, let λ_f be the linear functional on $j_L J(V)$ given by

$$T_L \longmapsto \lambda_f(T_L)$$

where for all $T \in J(V)$, we have $\lambda_f(T_L) = T(f)$.

Suppose $T \in J(V)$ and $\lambda_f(T_L) = 0$ for all $f \in C_c(\mathfrak{g}/L)$. Then $T_L = 0$. Since $\dim j_L J(V) < \infty$, this shows that the mapping

$$C_c(\mathfrak{g}/L) \longrightarrow \left(j_L J(V)\right)^*$$

with $f \mapsto \lambda_f$ is surjective. So we can choose $f \in C_c(\mathfrak{g}/L)$ such that $\lambda = \lambda_f$. Then

$$\mu_{\mathcal{O}}(f) = \lambda_f(\mu_{\mathcal{O},L}) = \lambda(\mu_{\mathcal{O},L}) = 0$$

for every regular orbit in V. Hence, by Lemma 4.1, $T(f) = 0$ for all $T \in J(V)$. This shows that $\lambda = \lambda_f = 0$. □

On the other hand we have the following simple lemma.

LEMMA 4.3 (Lemma 10). *Let ω be a compact set in $\mathfrak{g}$. Then we can choose another compact set ω_0 and a G-domain D in $\mathfrak{g}$ such that*

$$\omega_0^G \supset D \supset \omega.$$

PROOF. Let $\mathfrak{h}_1, \mathfrak{h}_2, \dots, \mathfrak{h}_r$ be a complete set of Cartan subalgebras of $\mathfrak{g}$ no two of which are conjugate under G. Then $\omega_i = \mathrm{cl}(\omega^G) \cap \mathfrak{h}_i$ is a compact subset of $\mathfrak{h}_i$. Let ω_0 be a compact neighborhood of $\cup_{i=1}^r \omega_i$ in $\mathfrak{g}$.

Choose $X \in \omega$. Define the G-domain D_X as follows. Let $\mathcal{O} = X^G$. Since $\mathrm{cl}(\mathcal{O})$ contains semisimple elements, we have $\mathrm{cl}(\mathcal{O}) \cap \omega_i \neq \emptyset$ for some i. Fix such an index i and choose $Y \in \mathrm{cl}(\mathcal{O}) \cap \omega_i$. Then ω_0 is a neighborhood of Y in $\mathfrak{g}$ and so, from Lemma 2.6, there exists a G-domain D_X in $\mathfrak{g}$ such that

$$Y \in D_X \subset \omega_0^G.$$

Then $D_X \cap \mathcal{O} \neq \emptyset$, and so $X \in \mathcal{O} \subset D_X$.

Now the sets D_X for $X \in \omega$ form an open covering of ω. Since ω is compact, we can choose a finite number of them $D_1, D_2 \dots, D_s$ such that $\omega \subset \cup_{1 \leq i \leq s} D_i = D \subset \omega_0^G$. □

We can now prove the first two parts of Theorem 4.4.

THEOREM 4.4 (Theorem 3). *Let ω be a compact subset of $\mathfrak{g}$ and $T \in J(\omega)$. Then there exists a locally summable function F on $\mathfrak{g}$ such that*

(1) $\widehat{T}(f) = \int_{\mathfrak{g}} F(X) \cdot f(X)\, dX$ *for all $f \in \mathcal{D}$,*
(2) F *is locally constant on $\mathfrak{g}'$, and*
(3) $|\eta|^{1/2} \cdot F$ *is locally bounded on $\mathfrak{g}$.*

PROOF. This proof does not prove (3).

Define D as in Lemma 4.3. Then $J(\omega) \subset J(D)$. Fix an open compact subset U of $\mathfrak{g}$. We can choose a lattice L in $\mathfrak{g}$ such that $\hat{f} \in C_c(\mathfrak{g}/L)$ for every $f \in C_c^\infty(U)$. From Corollary 4.2 we can choose a finite number of regular orbits $\mathcal{O}_1, \mathcal{O}_2, \dots, \mathcal{O}_r$ in D such that the $j_L \mu_i = \mu_{\mathcal{O}_i, L}$ span $j_L J(D)$.

Now fix $T \in J(D)$. Then there exist $c_i \in \mathbb{C}$ such that $T_L = \sum_{i=1}^r c_i \mu_{\mathcal{O}_i, L}$. Let $f \in C_c^\infty(U)$, then

$$\widehat{T}(f) = T(\hat{f}) = T_L(\hat{f}) = \sum_i c_i \mu_{\mathcal{O}_i}(\hat{f})$$

$$= \sum_i c_i \widehat{\mu_{\mathcal{O}_i}}(f).$$

This shows that $\widehat{T} = \sum_i c_i \widehat{\mu_{\mathcal{O}_i}}$ on U. Since the $\mathcal{O}_i$ are regular orbits, we know from the parts of Theorem 1.1 which we have proved that the $\widehat{\mu_{\mathcal{O}_i}}$ are locally summable functions on $\mathfrak{g}$ which are locally constant on $\mathfrak{g}'$. Hence $\widehat{T}$ is a locally summable function on U which is locally constant on $U \cap \mathfrak{g}'$. Since U was an arbitrary compact open subset of $\mathfrak{g}$, the assertion of the theorem is true for $\widehat{T}$. □

REMARK 4.5. Once Theorem 1.1 is proved completely, it would follow from the above proof that the function $|\eta|^{1/2} \cdot \widehat{T}$ is locally bounded on $\mathfrak{g}$.

5. Proof of Theorem 5.11

5.1. Proof of Lemma 5.1. Let γ be a semisimple element in $\mathfrak{g}$. Recall that $\mathcal{O}(\gamma)$ is the set of all G-orbits $\mathcal{O}$ in $\mathfrak{g}$ such that $\mathrm{cl}(\mathcal{O})$ contains γ. $\mathcal{O}(\gamma)$ is a finite set (see Corollary 2.11). Our first goal is to prove the following lemma.

LEMMA 5.1 (Lemma 11). *Suppose $f \in \mathcal{D}$ such that $\mu_{\mathcal{O}}(f) = 0$ for all $\mathcal{O} \in \mathcal{O}(\gamma)$. If $\mathfrak{h}$ is a Cartan subalgebra of $\mathfrak{g}$ containing γ, we can choose a neighborhood ω of γ in $\mathfrak{h}$ such that $\phi_f^{\mathfrak{h}} = 0$ on $\omega \cap \mathfrak{h}'$.*

Fix a Cartan subalgebra $\mathfrak{h}$ of $\mathfrak{g}$, and let Φ denote the space of all functions on $\mathfrak{h}'$ of the form $\phi = \phi_f^{\mathfrak{h}}$ for $f \in \mathcal{D}$. Given $\phi \in \Phi$, we write $\phi \sim 0$ if there exists a neighborhood ω of 0 in $\mathfrak{h}$ such that $\phi = 0$ on $\omega \cap \mathfrak{h}'$. Let Φ_0 denote the space of all $\phi \in \Phi$ such that $\phi \sim 0$. For any linear functional λ on Φ, put

$$T_\lambda(f) = \lambda(\phi_f)$$

for $f \in \mathcal{D}$. It is obvious that $T_\lambda \in J = J(\mathfrak{g})$.

LEMMA 5.2. *We have $T_\lambda \in J(\mathcal{N})$ if and only if $\lambda = 0$ on Φ_0.*

PROOF. "$\Rightarrow$" Suppose $T_\lambda \in J(\mathcal{N})$. Fix $f \in \mathcal{D}$ such that $\phi_f \sim 0$. Let ω be a neighborhood of 0 in $\mathfrak{h}$ such that $\phi_f = 0$ on $\omega \cap \mathfrak{h}'$. From Lemma 2.1 we can choose a G-domain D in $\mathfrak{g}$ such that $0 \in D \cap \mathfrak{h} \subset \omega$. Let F denote the characteristic function of D. Put $g = (1 - F) \cdot f$. Then $g = 0$ on D. Since $\mathcal{N} \subset D$, we have $T_\lambda(g) = 0$. Hence

$$T_\lambda(f) = T_\lambda(F \cdot f) = \lambda(\phi_{F \cdot f}).$$

But $\phi_{F \cdot f} = 0$ on $\mathfrak{h}' \cap {}^c D$. (The complement of a set S in the appropriate ambient set is denoted ${}^c S$.) Furthermore $\phi_{F \cdot f} = \phi_f = 0$ on $D \cap \mathfrak{h}'$. Therefore $\phi_{F \cdot f} = 0$. This shows that $T_\lambda(f) = \lambda(\phi_{F \cdot f}) = \lambda(0) = 0$. But $\lambda(\phi_f) = T_\lambda(f) = 0$. Consequently, $\lambda = 0$ on Φ_0.

"$\Leftarrow$" Suppose $\lambda = 0$ on Φ_0. Let f be an element in $\mathcal{D}$ such that $\mathrm{Supp}(f) \cap \mathcal{N} = \emptyset$. It is clear that $\phi_f \sim 0$. Therefore $T_\lambda(f) = \lambda(\phi_f) = 0$. $\square$

Let Λ denote the space of all linear functionals on Φ/Φ_0. Then any $\lambda \in \Lambda$ may also be regarded as a linear functional on Φ.

COROLLARY 5.3. *The map $\lambda \mapsto T_\lambda$ defines a linear injection of Λ into $J(\mathcal{N})$.*

PROOF. This is obvious. $\square$

COROLLARY 5.4. *We have $\dim(\Phi/\Phi_0) \leq [\mathcal{O}(0)]$.*

PROOF. Recall from Lemma 3.3 that the measures $\mu_{\mathcal{O}}$ form a base for $J(\mathcal{N})$. Since $\dim \Lambda \leq \dim J(\mathcal{N}) < \infty$ from the previous corollary, we have

$$\dim(\Phi/\Phi_0) = \dim \Lambda \leq \dim J(\mathcal{N}) = [\mathcal{O}(0)]. \quad \square$$

LEMMA 5.5. *Let f be an element in $\mathcal{D}$ such that $\mu_{\mathcal{O}}(f) = 0$ for all $\mathcal{O} \in \mathcal{O}(0)$. Then $\phi_f^{\mathfrak{h}} \sim 0$ for all Cartan subalgebras $\mathfrak{h}$ of $\mathfrak{g}$.*

PROOF. Fix $\mathfrak{h}$. If $\lambda \in \Lambda$, then $T_\lambda(f) = 0$ since $T_\lambda \in J(\mathcal{N})$. Hence $\lambda(\phi_f^{\mathfrak{h}}) = 0$. Since this is true for all $\lambda \in \Lambda$, we have $\phi_f^{\mathfrak{h}} \in \Phi_0$. Hence $\phi_f^{\mathfrak{h}} \sim 0$. $\qquad\square$

We are now prepared to prove Lemma 5.1.

PROOF. Let $\mathfrak{m}$ and M be the centralizers of γ in $\mathfrak{g}$ and G, respectively. By Corollary 2.3 we can fix a neighborhood U of γ in $\mathfrak{m}$ with the following properties.

(1) $\left|\eta_{\mathfrak{g}/\mathfrak{m}}(u)\right| \neq 0$ for all $u \in U$.
(2) Given any compact subset Q of $\mathfrak{g}$, there exists a compact set C in G such that $U^x \cap Q = \emptyset$ unless $x \in CM$.

Now fix f as in the statement of the lemma. Take $Q = \mathrm{Supp}(f)$ and choose C as above. We may select $\alpha \in C_c^\infty(G)$ such that if

$$\bar{\alpha}(\bar{x}) = \int_M \alpha(xm)\, dm$$

for $x \in G$ then $\bar{\alpha} = 1$ on $\bar{C}$ ($x \mapsto \bar{x}$ is the natural projection $G \to G/M = \bar{G}$). For $Y \in \mathfrak{m}$, put

$$g(Y) = \int_G \alpha(x) \cdot f(x.Y)\, dx.$$

Then $g \in C_c^\infty(\mathfrak{m})$.

Let $\mathfrak{h}$ be a Cartan subalgebra of $\mathfrak{g}$ containing γ. As usual, let A be the split component of the Cartan subgroup corresponding to $\mathfrak{h}$. Then, if $H \in U \cap \mathfrak{h}'$,

$$\int_{M/A} g(m.H)\, dm^* = \int_{M/A} dm^* \int_G \alpha(x) \cdot f(xm.H)\, dx$$

$$= \int_G \alpha(x)\, dx \int_{M/A} f(xm.H)\, dm^*$$

$$= \int_{G/M} d\bar{x} \int_M \alpha(\bar{x}m')\, dm' \int_{M/A} f(\bar{x}m'm.H)\, dm^*$$

$$= \int_{G/M} \bar{\alpha}(\bar{x})\, d\bar{x} \int_{M/A} f(\bar{x}m.H)\, dm^*.$$

But it follows from the definition of C that

$$\int_{M/A} f(xm.H)\, dm^* = 0$$

if $\bar{x} \notin \bar{C}$. On the other hand $\bar{\alpha}(\bar{x}) = 1$ for $\bar{x} \in \bar{C}$, hence

$$\int_{M/A} g(m.H)\, dm^* = \int_{G/M} d\bar{x} \int_{M/A} f(xm.H)\, dm^*$$

$$= \int_{G/A} f(x.H)\, dx^*.$$

This shows that

$$\phi_f^{G/\mathfrak{h}}(H) = \left|\eta_{\mathfrak{g}/\mathfrak{m}}(H)\right|^{1/2} \cdot \phi_g^{M/\mathfrak{h}}(H)$$

for $H \in U \cap \mathfrak{h}'$.

Define $g_\gamma \in C_c^\infty(\mathfrak{m})$ by $g_\gamma(Y) = g(\gamma + Y)$. Then it would be enough to show that $\phi_{g_\gamma}^{M/\mathfrak{h}} \sim 0$. For this it is enough to verify that if $\mathcal{O}_0$ is any nilpotent M-orbit in $\mathfrak{m}$ and if ν is the corresponding M-invariant measure on $\mathfrak{m}$, then $\nu(g_\gamma) = 0$.

Now $\mathcal{O}_0 = Y_0^M$ where Y_0 is a nilpotent element in $\mathfrak{m}$ which we may assume to be sufficiently near 0 so that $X_0 = \gamma + Y_0 \in U$. Put $\mathcal{O} = X_0^G$. By Lemma 2.8 we know that $\mathcal{O} \in \mathcal{O}(\gamma)$. For $x \in C_G(X_0)$,

$$(\gamma + Y_0)^x = \gamma + Y_0.$$

Since γ and Y_0 commute, γ is the semisimple component of X_0. Hence $\gamma^x = \gamma$ and $Y_0^x = Y_0$. This shows that $C_G(X_0) = C_M(Y_0)$. Now

$$\mu_{\mathcal{O}}(f) = \int_{G/C_G(X_0)} f(x.X_0)\, dx^* = \int_{G/M} d\bar{x} \int_{M/C_M(Y_0)} f(xm.X_0)\, dm^*.$$

On the other hand,

$$\nu(g_\gamma) = \int_{M/C_M(Y_0)} g_\gamma(m.Y_0)\, dm^* = \int_{M/C_M(Y_0)} g(m.X_0)\, dm^*$$

$$= \int_G dx \int_{M/C_M(Y_0)} \alpha(x) \cdot f(xm.X_0)\, dm^*$$

$$= \int_{G/M} \bar{\alpha}(\bar{x})\, d\bar{x} \int_{M/C_M(Y_0)} f(xm.X_0)\, dm^*$$

$$= \int_{G/M} d\bar{x} \int_{M/C_M(Y_0)} f(xm.X_0)\, dm^*$$

since $X_0 \in U$. Hence

$$\nu(g_\gamma) = \mu_{\mathcal{O}}(f) = 0$$

by our hypothesis on f. $\qquad\square$

5.2. Geometry of G-domains. Fix a compact neighborhood ω_0 of γ in $\mathfrak{g}$.

LEMMA 5.6 (Lemma 12). *We can choose G-domains V_i with $i \geq 1$ containing γ such that*

$$\omega_0^G \supset V_1 \supset V_2 \supset \cdots$$

and

$$\bigcap_{i \geq 1} V_i = \bigcup_{\mathcal{O} \in \mathcal{O}(\gamma)} \mathcal{O}.$$

PROOF. Let $\{\omega_i\}$ be a neighborhood basis of compact open subsets of γ such that $\gamma \in \cdots \subset \omega_2 \subset \omega_1 \subset \omega_0$. Let $V_0 = \omega_0^G$. As in Lemma 2.6, for $i \geq 1$ let V_i be a G-domain for the set $(V_{i-1} \cap \omega_i)$. Then $V_i \subset (V_{i-1} \cap \omega_i)^G = V_{i-1} \cap \omega_i^G \subset V_{i-1}$.

"$\subset$" If $X \notin \cup_{\mathcal{O} \in \mathcal{O}(\gamma)}\mathcal{O}$, then there exists $\ell > 0$ such that $\omega_\ell \cap X^G = \emptyset$. Therefore $X \notin \omega_\ell^G$, and so $X \notin V_\ell$.

"$\supset$" If $X \in \cup_{\mathcal{O} \in \mathcal{O}(\gamma)}\mathcal{O}$, then for all $i \geq 1$, we have $V_i \cap X^G \neq \emptyset$. Therefore $X \in V_i$ for all $i \geq 1$. $\qquad\square$

Let L be a lattice in $\mathfrak{g}$.

COROLLARY 5.7. *Let $\{V_i\}$ be a sequence of G-domains as above. Then there exists an index i_0 such that $j_L J(V_i)$ is spanned by the $j_L \mu_{\mathcal{O}}$ with $\mathcal{O} \in \mathcal{O}(\gamma)$ for all $i \geq i_0$.*

PROOF. By Theorem 12.1, $\dim j_L(\omega_0) < \infty$. So we can choose i_0 such that $j_L J(V_i) = j_L J(V_{i_0})$ for all $i \geq i_0$. Let λ be a linear functional on $j_L J(\omega_0)$ such that $\lambda(\mu_{\mathcal{O},L}) = 0$ for all $\mathcal{O} \in \mathcal{O}(\gamma)$. It is enough to verify that $\lambda = 0$ on $j_L J(V_{i_0})$.

Again by Theorem 12.1 there exists $f \in C_c(\mathfrak{g}/L)$ such that $\lambda(T_L) = T(f)$ for all $T \in J(\omega_0)$. Suppose $\lambda \neq 0$ on $j_L J(V_{i_0})$. Then, by Corollary 4.2, we can choose, for every $i \geq i_0$, a regular orbit $\mathcal{O}_i$ in V_i such that $\mu_{\mathcal{O}_i}(f) \neq 0$. Since $V_i \subset \omega_0^G$, we can arrange, by selecting a subsequence, that the following condition holds. There exists a Cartan subalgebra $\mathfrak{h}$ of $\mathfrak{g}$ and elements $H_0 \in \mathfrak{h}$ and $H_i \in \mathfrak{h}' \cap V_i$ such that

$$\phi_f^{\mathfrak{h}}(H_i) \neq 0$$

and H_i converges to H_0 in $\mathfrak{h}$ as $i \to \infty$. From Lemma 5.1 we have that there exists $\mathcal{O} \in \mathcal{O}(H_0)$ such that $\mu_{\mathcal{O}}(f) \neq 0$. Since $H_i \in V_i \subset V_j$ for $i \geq j$ we have $H_0 \in \mathrm{cl}(V_i) = \overline{V_i}$ for all i. This implies that $H_0 \in \cap V_i = \cup_{\mathcal{O} \in \mathcal{O}(\gamma)} \mathcal{O}$. Since V_i is open and $H_0 \in \mathrm{cl}(\mathcal{O})$, it follows that $\mathcal{O} \subset V_i$. Since this is true for all i, $\mathcal{O} \subset \cap V_i = \cup_{\mathcal{O} \in \mathcal{O}(\gamma)} \mathcal{O}$. So

$$0 \neq \mu_{\mathcal{O}}(f) = \lambda(\mu_{\mathcal{O},L}) = 0$$

giving a contradiction. $\qquad\qquad\qquad\qquad\qquad\qquad\qquad\qquad\qquad\qquad\qquad\square$

5.3. Proof of Theorem 5.11. We shall now follow [23] in our proof of Theorem 5.11.

LEMMA 5.8 (Lemma 13). *Let ω be a compact subset of $\mathfrak{g}$. Then we can choose a lattice L in $\mathfrak{g}$ such that $j_L J(\omega)$ is contained in the space spanned by the $\mu_{\mathcal{O},L}$ with $\mathcal{O} \in \mathcal{O}(0)$.*

PROOF. Fix a lattice $L_0 \supset \omega$. It will be enough to prove the lemma for L_0 instead of ω. Put $R' = R \cap \Omega^{\times}$. Choose $t_1, t_2 \in R'$ such that $|t_1| \geq |t_2| > 0$. Since $t_2 L_0 \subset t_1 L_0 \subset L_0$, it is obvious from Theorem 12.1 that we can choose $t_0 \in R'$ such that

$$j_{L_0} J(t_0^2 L_0) = \bigcap_{t \in R'} j_{L_0} J(t L_0).$$

By Lemma 2.6 we can choose a G-domain V in $\mathfrak{g}$ such that $0 \in V \subset L_0^G$. Since $t_1 L_0 \subset V$ for some $t_1 \in R'$, we have $J(t t_1 L_0) \subset J(t V) \subset J(t L_0)$, and so

$$\bigcap_{t \in R'} j_{L_0} J(t L_0) = \bigcap_{t \in R'} j_{L_0} J(t V).$$

Hence we conclude from Corollary 5.7 that $j_{L_0} J(t_0^2 L_0)$ is spanned by $j_{L_0} \mu_{\mathcal{O}} = \mu_{\mathcal{O},L_0}$ where $\mathcal{O}$ runs over all orbits contained in

$$\bigcap_{t \in R'} t V = \mathcal{N}.$$

Now let $T \in J(L_0)$. In the notation of Lemma 3.2 we have

$$\langle \rho(t_0^{-2})T, f_{t_0^2} \rangle = \langle T, f \rangle$$

and so $\rho(t_0^{-2})T \in J(t_0^2 L_0)$. Hence we can choose constants $c_\mathcal{O} \in \mathbb{C}$ such that

$$j_{L_0}\left[\rho(t_0^{-2})T - \sum_\mathcal{O} c_\mathcal{O} \cdot \mu_\mathcal{O} \right] = 0.$$

Put $L = t_0^{-2} L_0$. Then the above relation implies that

$$j_L\left[T - \sum_\mathcal{O} c_\mathcal{O} \cdot \rho(t_0^2)\mu_\mathcal{O} \right] = 0.$$

The desired result now follows from Lemma 3.2. $\qquad\square$

LEMMA 5.9 (Lemma 14). *For any lattice L in $\mathfrak{g}$, the $\mu_{\mathcal{O},L}$ with $\mathcal{O} \in \mathcal{O}(0)$ are linearly independent.*

PROOF. Let $\mathcal{O}_1, \mathcal{O}_2, \ldots, \mathcal{O}_s$ be all the distinct nilpotent G-orbits. Put $\mu_i = \mu_{\mathcal{O}_i}$. It follows from Lemma 3.3 that we can choose $f_j \in \mathcal{D}$ such that $\mu_i(f_j) = \delta_{ij}$. Fix a lattice L_0 in $\mathfrak{g}$ such that $f_j \in C_c(\mathfrak{g}/L_0)$ for $1 \leq j \leq s$. Choose $t \in \Omega^\times$ such that $L \subset t^2 L_0$ and put $g_j = (f_j)_{t^2}$. Then $g_j \in C_c(\mathfrak{g}/L)$, and we conclude from Lemma 3.2 that

$$\mu_i(g_j) = \mu_i\big((f_j)_{t^2}\big) = \big(\rho(t^2)\mu_i\big)(f_j) = |t|^{d(\mathcal{O}_i)} \delta_{ij}.$$

The lemma follows immediately. $\qquad\square$

COROLLARY 5.10. *Let V be an open neighborhood of 0 in $\mathfrak{g}$. Then the functions $\widehat{\mu_\mathcal{O}}$, indexed by $\mathcal{O} \in \mathcal{O}(0)$, are linearly independent on $V \cap \mathfrak{g}'$.*

PROOF. This follows immediately from Lemma 5.9. $\qquad\square$

THEOREM 5.11 (Theorem 4). *Let ω be a compact subset of $\mathfrak{g}$. Then there exists a G-domain D containing 0 with the following property. For every $T \in J(\omega)$ we can choose complex numbers $c_\mathcal{O}(T)$ such that*

$$\widehat{T} = \sum_{\mathcal{O} \in \mathcal{O}(0)} c_\mathcal{O}(T) \cdot \widehat{\mu_\mathcal{O}}$$

on D. Moreover, if V is any neighborhood of 0 in $\mathfrak{g}$, the functions $\widehat{\mu_\mathcal{O}}$, indexed by $\mathcal{O} \in \mathcal{O}(0)$, are linearly independent on $V \cap \mathfrak{g}'$.

PROOF. Fix a lattice $L_0 \supset \omega$. It will be enough to prove the theorem for L_0 instead of ω. Choose L as in Lemma 5.8, and fix a compact open neighborhood U of 0 in $\mathfrak{g}$ such that $\hat{f} \in C_c(\mathfrak{g}/L)$ for every $f \in C_c^\infty(U)$. Choose $T \in J(L_0)$. By Lemma 5.8 we can choose complex numbers $c_\mathcal{O}$ such that

$$j_L T = T_L = \sum_{\mathcal{O} \in \mathcal{O}(0)} c_\mathcal{O} \cdot \mu_{\mathcal{O},L}.$$

Then, for $f \in C_c^\infty(U)$,

$$\widehat{T}(f) = T(\hat{f}) = T_L(\hat{f}) = \sum_{\mathcal{O} \in \mathcal{O}(0)} c_\mathcal{O} \cdot \mu_{\mathcal{O}, L}(\hat{f})$$

$$= \sum_{\mathcal{O} \in \mathcal{O}(0)} c_\mathcal{O} \cdot \widehat{\mu_\mathcal{O}}(f).$$

Now, by Lemma 2.6, we can choose a G-domain D such that $0 \in D \subset U^G$. The remaining assertion follows from Corollary 5.10. □

5.4. A translation of Theorem 5.11. Fix $Z \in \mathfrak{z}$ and define, for all $X \in \mathfrak{g}$,

$$\lambda_Z(X) = \chi\big(B(Z, X)\big).$$

Then λ_Z is a G-invariant function in $C^\infty(\mathfrak{g})$ and $\lambda_Z = 1$ on $\mathfrak{g}_1 = [\mathfrak{g}, \mathfrak{g}]$. Moreover, $\lambda_Z T \in J(\omega)$ if $T \in J(\omega)$, and $(\lambda_Z T)\widehat{\ }(X) = \widehat{T}(X + Z)$ for $X \in \mathfrak{g}$ and $T \in J(\omega)$.

LEMMA 5.12. *Let σ be a distribution on $\mathfrak{g}$ such that $\mathrm{Supp}(\sigma) \subset \mathfrak{g}_1$. Then for $Z \in \mathfrak{z}$,*

$$(\lambda_Z \sigma)\widehat{\ } = \widehat{\sigma}.$$

PROOF. Since $\lambda_Z = 1$ on $\mathfrak{g}_1$, we have $\lambda_Z \sigma = \sigma$. □

COROLLARY 5.13. *Let ω be a compact subset of $\mathfrak{g}$. Fix $Z \in \mathfrak{z}$. Then there exists a G-domain D containing Z satisfying the following condition. For any $T \in J(\omega)$, we can choose complex numbers $c_\mathcal{O}(T)$ such that*

$$\widehat{T} = \sum_{\mathcal{O} \in \mathcal{O}(0)} c_\mathcal{O}(T) \cdot \widehat{\mu_\mathcal{O}}$$

on D.

PROOF. Choose a G-domain D_0 containing 0 and satisfying the conditions of Theorem 5.11. Fix $T \in J(\omega)$. Then $\lambda_Z T \in J(\omega)$. Hence, we can choose constants $c_\mathcal{O} \in \mathbb{C}$ such that

$$(\lambda_Z T)\widehat{\ } = \sum_{\mathcal{O} \in \mathcal{O}(0)} c_\mathcal{O} \cdot \widehat{\mu_\mathcal{O}}$$

on D_0.

Now $\mathrm{Supp}(\mu_\mathcal{O}) \subset \mathcal{N} \subset \mathfrak{g}_1$. Hence $(\lambda_Z \mu_\mathcal{O})\widehat{\ } = \widehat{\mu_\mathcal{O}}$. Therefore, for $X \in \mathfrak{g}$,

$$\widehat{\mu_\mathcal{O}}(X + Z) = \widehat{\mu_\mathcal{O}}(X).$$

Hence, for $X \in D_0$,

$$\widehat{T}(X + Z) = \sum_{\mathcal{O} \in \mathcal{O}(0)} c_\mathcal{O} \cdot \widehat{\mu_\mathcal{O}}(X + Z).$$

Let $D = D_0 + Z$. □

6. Application of the induction hypothesis

In order to complete the proof of Theorem 1.1, it remains to verify the following result.

THEOREM 6.1 (Theorem 11). *Let $\mathcal{O}$ be a regular orbit in $\mathfrak{g}$. Then the function $|\eta|^{1/2} \cdot \widehat{\mu_{\mathcal{O}}}$ is locally bounded on $\mathfrak{g}$.*

We have already proved that $\widehat{\mu_{\mathcal{O}}}$ is a locally summable function on $\mathfrak{g}$ that is locally constant on $\mathfrak{g}'$. Assuming Theorem 6.1, we recall some of its consequences.

COROLLARY 6.2. *Let ω be a compact subset of $\mathfrak{g}$. Then, for all $T \in J(\omega)$, the function $|\eta|^{1/2} \cdot \widehat{T}$ is locally bounded on $\mathfrak{g}$.*

PROOF. See Remark 4.5.   □

COROLLARY 6.3. *Let $\mathcal{O}$ be any orbit in $\mathfrak{g}$. Then the function $|\eta|^{1/2} \cdot \widehat{\mu_{\mathcal{O}}}$ is locally bounded on $\mathfrak{g}$.*

PROOF. This is a special case of Corollary 6.2.   □

6.1. Recasting Theorem 6.1. Throughout the remainder of this section and §7 we will assume that Theorem 6.1 is true for all appropriate objects with dimension less than $\dim G$.

Let $\mathfrak{g}_e$ denote the set of all $X \in \mathfrak{g}'$ such that $C_{\mathfrak{g}}(X)$ is an elliptic Cartan subalgebra. Then $\mathfrak{g}_e$ is an open, G-invariant subset of $\mathfrak{g}'$. A regular G-orbit $\mathcal{O}$ is called elliptic if $\mathcal{O} \subset \mathfrak{g}_e$.

LEMMA 6.4 (Lemma 15). *Suppose $\mathcal{O}$ is a regular G-orbit in $\mathfrak{g}$ which is not elliptic. Then the function $|\eta|^{1/2} \cdot \widehat{\mu_{\mathcal{O}}}$ is locally bounded on $\mathfrak{g}$.*

PROOF. See Corollary 1.14 and the induction hypothesis of Theorem 6.1.   □

Fix an element $\theta \in \mathcal{D}$ such that, for all $f \in \mathcal{D}$,

$$\int_{G/Z} dx^* \left| \int_{\mathfrak{g}} f(X) \cdot \theta(x.X) \, dX \right| < \infty.$$

(See Lemma 1.15 for the existence of such a nonzero θ.)

THEOREM 6.5 (Theorem 12). *Define $\Theta \in J$ by*

$$\Theta(f) = \int_{G/Z} dx^* \int_{\mathfrak{g}} f(X) \cdot \theta(x.X) \, dX.$$

then Θ is a locally summable function on $\mathfrak{g}$ which is locally constant on $\mathfrak{g}'$. Moreover, $|\eta|^{1/2} \cdot \Theta$ is locally bounded on $\mathfrak{g}$.

Note that

$$\widehat{\Theta}(f) = \Theta(\hat{f}) = \int_{G/Z} dx^* \int_{\mathfrak{g}} \hat{f}(X) \cdot \theta(x.X) \, dX$$

$$= \int_{G/Z} dx^* \int_{\mathfrak{g}} f(X) \cdot \widehat{\theta}(x.X) \, dX.$$

Hence it is clear that $\widehat{\Theta} \in J\big(\mathrm{Supp}(\widehat{\theta})\big)$. Since $\mathrm{Supp}(\widehat{\theta})$ is compact, it follows from Theorem 4.4 that Θ is a locally summable function on $\mathfrak{g}$ which is locally constant on $\mathfrak{g}'$. So it remains to prove that $|\eta|^{1/2} \cdot \Theta$ is locally bounded.

LEMMA 6.6 (Lemma 16). *Theorem 6.1 and Theorem 6.5 are equivalent.*

PROOF. Theorem 6.5 is an immediate consequence of Corollary 6.2.

Conversely, from Lemma 6.4 we need only deal with regular elliptic orbits. From Lemma 1.19 it follows that for all compact $U \subset \mathfrak{g}$ there exists $\theta \in C_c^\infty(\mathfrak{g}_\mathfrak{h})$ such that $\Theta(f) = \phi_{\hat{f}}(H_0)$ for all $f \in C_c^\infty(U)$. Consequently, if $|\eta|^{1/2} \cdot \Theta$ is locally bounded on $\mathfrak{g}$, then $|\eta|^{1/2} \cdot \widehat{\mu_{\mathcal{O}}}$ is locally bounded on $\mathfrak{g}$. $\qquad\square$

6.2. First step in the proof of Theorem 6.1. From Lemma 6.6 we can assume that both Theorem 6.5 and Theorem 6.1 hold for all appropriate objects with dimension less than $\dim G$. In this section, we show that if $\mathcal{O}$ is a regular G-orbit in $\mathfrak{g}$ and γ is a non-central, semisimple element of $\mathfrak{g}$, then $|\eta|^{1/2} \cdot \widehat{\mu_{\mathcal{O}}}$ is bounded on some neighborhood of γ.

Let $\mathfrak{z}$ be the center of $\mathfrak{g}$. Fix a semisimple element $\gamma \notin \mathfrak{z}$ in $\mathfrak{g}$. As before, let $M = C_G(\gamma)$ and $\mathfrak{m} = C_\mathfrak{g}(\gamma)$. Fix a compact neighborhood ω of γ in $\mathfrak{m}$, and choose an M-domain U in $\mathfrak{m}$ satisfying the conditions of Corollary 2.2 and Corollary 2.3. In particular, $\gamma \in U \subset \omega^M$. Put $V = U^G$. Then V is a G-domain in $\mathfrak{g}$. Define an M-invariant neighborhood U_0 of 0 in $\mathfrak{m}$ by $U = U_0 + \gamma$. Consider the mapping

$$(x, \gamma) \mapsto (\gamma + Y)^x$$

of $G \times U_0$ into V. It is surjective. Since $\det\big(\mathrm{ad}(\gamma + Y)|_{\mathfrak{g}/\mathfrak{m}}\big) \neq 0$ for $Y \in U_0$, it is everywhere submersive. Hence, following the proof of Lemma 3.9, there exists a surjective, linear map $\alpha \mapsto f_\alpha$ of $C_c^\infty(G \times U_0)$ into $C_c^\infty(V)$ with the following property. If $F \in C(V)$, then

$$\int_{G \times U_0} \alpha(x : Y) \cdot F\big((\gamma + Y)^x\big)\, dx\, dY = \int_V f_\alpha(X) \cdot F(X)\, dX.$$

Also, there exists an M-invariant distribution τ_0 on U_0 such that

$$\Theta(f_\alpha) = \tau_0(\beta_\alpha)$$

for all $\alpha \in C_c^\infty(G \times U_0)$. Here β_α is the element of $C_c^\infty(U_0)$ given by

$$\beta_\alpha(Y) = \int_G \alpha(x : Y)\, dx.$$

Here Θ has the same meaning as in Theorem 6.5. The mapping $\alpha \mapsto \beta_\alpha$ of $C_c^\infty(G \times U_0)$ into $C_c^\infty(U_0)$ is obviously surjective. Let F_0 be the characteristic function of U_0. For all $f \in C_c^\infty(\mathfrak{m})$, define $\tau(f) = \tau_0(F_0 \cdot f)$. Then τ is an M-invariant distribution whose support lies in U_0. Let K_0 be a compact open subgroup of G such that $\theta^k = \theta$ for all $k \in K_0$. Let dk be the normalized measure on K_0. Then

$$\tau(\beta) = \int_{G/Z} dx^* \int_{\mathfrak{m} \times K_0} F_0(Y) \cdot \beta(Y) \cdot \theta\big(xk.(\gamma + Y)\big)\, dY\, dk$$

for $\beta \in C_c^\infty(\mathfrak{m})$.

By condition (2) of Corollary 2.3 we can choose a compact set C in G such that

$$(\gamma + U_0)^x \cap \operatorname{Supp}(\theta) = \emptyset$$

for $x \in G$ unless $x \in CM$. Choose $y_1, y_2, \ldots, y_r \in G$ such that $C \subset \cup_{i=1}^r K_0 y_i$. Put

$$h_x(Y) = F_0(Y)\theta\big(x.(\gamma + Y)\big)$$

for $Y \in \mathfrak{m}$ and $x \in G$. Since U_0 is closed (and open) and $\operatorname{Supp}(\theta)$ is compact, $h_x \in C_c^\infty(\mathfrak{m})$.

LEMMA 6.7. *We have $h_x = 0$ unless $x \in \cup_{i=1}^r K_0 y_i M$.*

PROOF. Suppose $h_x \neq 0$. Then we can choose $Y \in U_0$ such that $\theta\big(x.(\gamma+Y)\big) \neq 0$. Then $(\gamma + U_0)^x \cap \operatorname{Supp}(\theta) \neq \emptyset$. Hence $x \in CM \subset \cup_i K_0 y_i M$. □

Put $h_i = h_{y_i}$ for $1 \leq i \leq r$. If $x = k y_i m^{-1}$, it follows that, for $Y \in \mathfrak{m}$,

$$h_x(Y) = F_0(Y) \cdot \theta\big(k y_i m^{-1}.(\gamma + Y)\big) = F_0(Y) \cdot \theta^{k^{-1}}\big(y_i.(\gamma + m^{-1}.Y)\big)$$
$$= F_0(Y) \cdot \theta\big(y_i.(\gamma + m^{-1}.Y)\big) = (h_i)^m(Y).$$

LEMMA 6.8. *Let $\widehat{h_x}$ denote the Fourier transform of h_x on $\mathfrak{m}$, and let $\omega_0 = \cup_{i=1}^r \operatorname{Supp}(\widehat{h_i})$. Then $\operatorname{Supp}(\widehat{\tau}) \subset \operatorname{cl}(\omega_0^M)$.*

PROOF. Note that, for $\beta \in C_c^\infty(\mathfrak{m})$,

$$\widehat{\tau}(\beta) = \tau(\widehat{\beta})$$
$$= \int_{G/Z} dx^* \int_{\mathfrak{m}} dY \int_{K_0} F_0(Y) \cdot \widehat{\beta}(Y) \cdot \theta\big(xk.(\gamma + Y)\big)\, dk$$
$$= \int_{G/Z} dx^* \int_{\mathfrak{m}} dY \int_{K_0} h_{xk}(Y) \cdot \widehat{\beta}(Y)\, dk$$
$$= \int_{G/Z} dx^* \int_{K_0} dk \int_{\mathfrak{m}} \beta(Y) \cdot \widehat{h_{xk}}(Y)\, dY.$$

Suppose $\beta = 0$ on ω_0^M. If $\widehat{\tau}(\beta) \neq 0$, then $\beta(Y) \cdot \widehat{h_x}(Y) \neq 0$ for some $x \in G$ and $Y \in \mathfrak{m}$. Therefore $h_x \neq 0$, and, from the above lemma, $x = k y_i m^{-1}$. But then $h_x = (h_i)^m$, and therefore

$$\widehat{h_x} = (\widehat{h_i})^m.$$

Hence

$$Y \in \operatorname{Supp}(\widehat{h_x}) = \big(\operatorname{Supp}(\widehat{h_i})\big)^m \subset \omega_0^M.$$

But since $\beta = 0$ on ω_0^M, we conclude that $\beta(Y) \cdot \widehat{h_x}(Y) = 0$. This is a contradiction. □

Now we can apply Theorem 5.11 to τ and choose an M-domain U_1 in $\mathfrak{m}$ such that $0 \in U_1 \subset U_0$ and

$$\tau = \sum_\xi c_\xi \cdot \widehat{\nu_\xi}$$

on U_1. Here $c_\xi \in \mathbb{C}$, and ξ runs over all nilpotent M-orbits in $\mathfrak{m}$. Also ν_ξ is the corresponding M-invariant measure on $\mathfrak{m}$, and $\widehat{\nu}_\xi$ is its Fourier transform on $\mathfrak{m}$.

By replacing U_0 by U_1 we may assume that $U_1 = U_0$.

LEMMA 6.9. *There exists an M-domain U_0 in $\mathfrak{m}$ containing 0 and complex numbers c_ξ such that*

$$\int_\mathfrak{m} \beta(Y) \cdot \Theta(\gamma + Y)\, dY = \sum_\xi c_\xi \cdot \widehat{\nu}_\xi(\beta)$$

for all $\beta \in C_c^\infty(U_0)$. Here ξ runs over all nilpotent M-orbits in $\mathfrak{m}$.

PROOF. Choose $\beta \in C_c^\infty(U_0)$. Since $\alpha \mapsto \beta_\alpha$ is surjective, there exists $\alpha \in C_c^\infty(G \times U_0)$ such that $\beta_\alpha = \beta$. Then

$$\Theta(f_\alpha) = \int_\mathfrak{g} f_\alpha(X) \cdot \Theta(X)\, dX$$

$$= \int_{G \times U_0} \alpha(x : Y) \cdot \Theta(\gamma + Y)\, dx\, dY$$

$$= \int_{U_0} \beta(Y) \cdot \Theta(\gamma + Y)\, dY.$$

But $\tau(\beta) = \tau(\beta_\alpha) = \Theta(f_\alpha)$ and so the assertion is obvious. $\square$

COROLLARY 6.10. *We can choose an M-domain U in $\mathfrak{m}$ containing γ and complex numbers c_ξ such that*

$$\Theta = \sum_\xi c_\xi \cdot \widehat{\nu}_\xi$$

on U as functions. Here ξ runs over all nilpotent M-orbits in $\mathfrak{m}$.

PROOF. This is immediate by translation by γ (see Corollary 5.13). $\square$

COROLLARY 6.11. *The function $|\eta_\mathfrak{m}|^{1/2} \cdot \Theta$ remains bounded in some neighborhood of γ in $\mathfrak{m}$.*

PROOF. Since $\dim \mathfrak{m} < \dim \mathfrak{g}$, this follows from the induction hypothesis (see Corollary 6.3). $\square$

COROLLARY 6.12 (Lemma 17). *The function $|\eta|^{1/2} \cdot \Theta$ remains bounded on some neighborhood of γ in $\mathfrak{g}$.*

PROOF. This is obvious from Corollary 6.11 $\square$

COROLLARY 6.13. *Let $\mathcal{O}$ be a regular G-orbit in $\mathfrak{g}$. Then $|\eta|^{1/2} \cdot \widehat{\mu_\mathcal{O}}$ remains bounded on some neighborhood of γ in $\mathfrak{g}$.*

PROOF. In view of Lemma 6.4, we may assume that $\mathcal{O}$ is elliptic. The assertion then follows from Lemma 1.19 and Corollary 6.12. $\square$

7. Reformulation of the problem and completion of the proof

Recall that in §6 we showed that, for $\mathcal{O}$ regular, $|\eta|^{1/2} \cdot \widehat{\mu_{\mathcal{O}}}$ is bounded near any fixed semisimple, non-central element of $\mathfrak{g}$. We shall now approach our problem from a somewhat different angle.

Fix an open compact subgroup K of G and let dk denote the normalized Haar measure of K. Let $\mathfrak{h}_1$ and $\mathfrak{h}_2$ be two Cartan subalgebras of $\mathfrak{g}$, and let A_2 be the split component of the Cartan subgroup of G corresponding to $\mathfrak{h}_2$. Let $x \mapsto x^*$ denote the projection of G on $G^* = G/A_2$. Let dx^* be an invariant measure on G^*. Recall that χ is a non-trivial additive character of Ω.

LEMMA 7.1 (Lemma 18). *Fix compact subsets $\omega_1 \subset \mathfrak{h}_1'$ and $\omega_2 \subset \mathfrak{h}_2'$. Then we can choose a compact set C^* in G^* with the following property. Suppose $H_i \in \omega_i$ and $x \in G$. Then the integral*

$$\int_K \chi\big(B(k.H_1, x.H_2)\big) \, dk$$

is zero unless $x^* \in C^*$.

We need some preparation.

LEMMA 7.2. *Suppose $\mathfrak{g}$ is semisimple. Fix $X, Y \in \mathfrak{g}$ and suppose $B(X, [Z, Y])$ is zero for all $Z \in \mathfrak{g}$. Then $[X, Y] = 0$.*

PROOF.

$$0 = B(X, [Z, Y]) = B([X, Z], Y) = -B(Z, [X, Y]).$$

Since B is nondegenerate, this implies that $[X, Y] = 0$. $\square$

COROLLARY 7.3. *Suppose $\mathfrak{g}$ is semisimple. Consider the mapping*

$$x \mapsto B(X, x.Y)$$

from G to Ω. This map is submersive at x if and only if $[X, x.Y] \neq 0$.

PROOF. Since the differential of this map sends $Z \in \mathfrak{g}$ to $B(X, x.[Z, Y])$, the map is submersive if and only if $[x^{-1}.X, Y] \neq 0$ by Lemma 7.2. $\square$

For $X \in \mathfrak{g}$, let $|X|$ denote the norm of the matrix X as discussed in §2. Define the integer $\nu(X)$ by $|X| = q^{-\nu(X)}$ ($\nu(X) = \infty$ if $X = 0$). Fix $\varpi \in \Omega^\times$ such that $|\varpi| = q^{-1}$. Then $|X| = |\varpi^{\nu(X)}|$. Let S be the set of all $X \in \mathfrak{g}$ with $|X| = 1$. Then S is an open and compact subset of $\mathfrak{g}$.

LEMMA 7.4. *Suppose $\mathfrak{g}$ is semisimple. Let $\mathfrak{h}$ be a Cartan subalgebra of $\mathfrak{g}$ and ω a compact subset of $\mathfrak{h}'$. Let S_ω be the smallest closed subset of S containing the set*

$$\{\varpi^{-\nu(x.H)} \cdot x.H : x \in G \text{ and } H \in \omega\}.$$

Let $X \in S_\omega$. Then X is either regular or nilpotent. Moreover, it is regular if and only if it is of the form $\varpi^{-\nu(x.H)} \cdot x.H$ for some $x \in G$ and $H \in \omega$.

PROOF. Suppose $X \in S_\omega$. We can choose sequences $\{x_i\}$ in G and $\{H_i\}$ in ω such that $\varpi^{-\nu(x_i.H_i)} \cdot x_i.H_i \to X$.

First suppose $|x_i.H_i|$ remains bounded. Since ω is compact and contained in $\mathfrak{h}'$, we may, by choosing a subsequence, assume that $x_i.H_i \to x.H$ with $H \in \mathfrak{h}'$. Then $X = \varpi^{-\nu(x.H)} \cdot x.H$.

So now assume, by taking a subsequence, that $|x_i.H_i| \to \infty$. The eigenvalues of $\mathrm{ad}(x_i.H_i)$ are the same as those of $\mathrm{ad}(H_i)$ and so remain bounded. Hence, the eigenvalues of $\varpi^{-\nu(x_i.H_i)} \cdot \mathrm{ad}(x_i.H_i)$ tend to zero. Hence X is nilpotent. □

LEMMA 7.5. *Suppose $\mathfrak{g}$ is semisimple. Fix compact sets $\omega_1 \subset \mathfrak{h}'_1$ and $\omega_2 \subset \mathfrak{h}'_2$. Suppose $(H_1, X) \in \omega_1 \times S_{\omega_2}$ and X is nilpotent. Then the mapping*

$$k \mapsto B(H_1, k.X)$$

from K to Ω is everywhere submersive.

PROOF. Since H_1 is regular, its centralizer in $\mathfrak{g}$ is $\mathfrak{h}_1$. But $k.X$ is nilpotent and nonzero, hence $k.X \notin \mathfrak{h}_1$. Therefore $[H_1, k.X] \neq 0$. So we can apply Corollary 7.3. □

We now prove Lemma 7.1.

PROOF. Put $G_1 = G \cap \mathbf{G_1}$ where $\mathbf{G_1}$ is the "derived group" of $\mathbf{G}$. Then $G/G_1 A_2$ is compact. Hence, without loss of generality, we may assume that $\mathfrak{g}$ is semisimple.

We now assume, as we may, that ω_1 is open (as well as compact). Let S_2 be a compact, open neighborhood of $S_{\omega_2} \cap \mathcal{N}$ in S such that the mapping

$$(H_1, k, X) \mapsto (H_1, B(H_1, k.X), X)$$

of $\omega_1 \times K \times S_2$ into $\omega_1 \times \Omega \times S$ is everywhere submersive. So for $\alpha \in C_c^\infty(\omega_1 \times K \times S_2)$ there exists $f_\alpha \in C_c^\infty(\omega_1 \times \Omega \times S)$ such that

$$\int_{\omega_1 \times K \times S_2} \alpha(H_1 : k : X) \cdot \Phi(H_1 : B(H_1, k.X) : X) \, dk \, dH_1 \, dX$$

$$= \int_{\omega_1 \times \Omega \times S} f_\alpha(H_1 : t : X) \cdot \Phi(H_1 : t : X) \, dH_1 \, dt \, dX$$

for all $\Phi \in C^\infty(\omega_1 \times \Omega \times S)$. Now let $\alpha = 1$ and write $f = f_\alpha$ in this case. Then

$$\int_{\omega_1 \times K \times S_2} \Phi(H_1 : B(H_1, k.X) : X) \, dk \, dH_1 \, dX$$

$$= \int_{\omega_1 \times \Omega \times S} f(H_1 : t : X) \cdot \Phi(H_1 : t : X) \, dH_1 \, dt \, dX.$$

Taking Φ of the form

$$\Phi(H_1 : t : X) = \beta(H_1) \cdot \psi(t) \cdot \phi(X)$$

we easily deduce that

$$\int_K \psi\big(B(H_1, k.X)\big)\, dk = \int_\Omega f(H_1 : t : X) \cdot \psi(t)\, dt$$

for all $\psi \in C^\infty(\Omega)$ and $(H_1, X) \in \omega_1 \times S_2$.

Since $f \in C_c^\infty(\omega_1 \times \Omega \times S)$, there exists an integer $N \geq 0$ such that

$$f(H_1 : t + \lambda : X) = f(H_1 : t : X)$$

if $\lambda \in \varpi^N R$.

Now fix $t_0 \in \Omega^\times$. Then

$$\int_K \chi\big(t_0 B(H_1, k.X)\big)\, dk = \int_\Omega f(H_1 : t : X) \cdot \chi(t_0 t)\, dt$$

$$= \sum_{t \in \Omega/\varpi^N R} f(H_1 : t : X) \cdot \chi(t_0 t) \int_{\varpi^N R} \chi(t_0 \lambda)\, d\lambda.$$

But

$$\int_{\varpi^N R} \chi(t_0 \lambda)\, d\lambda = |t_0|^{-1} \int_{\varpi^{N+\nu_0} R} \chi(t)\, dt$$

where $\nu_0 = \nu(t_0)$. Let $\varpi^{\mathfrak{f}_0} R$ be the conductor of χ (i.e., $\varpi^{\mathfrak{f}_0} R \subset \ker \chi$, but $\varpi^{\mathfrak{f}_0-1} R \not\subset \ker \chi$). If $N + \nu_0 < \mathfrak{f}_0$, then

$$\int_{\varpi^{N+\nu_0} R} \chi(t)\, dt = 0.$$

Hence, if $N + \nu(t_0) < \mathfrak{f}_0$, then $\int_K \chi\big(t_0 B(H_1, k.X)\big)\, dk = 0$ for $(H_1, X) \in \omega_1 \times S_2$.

Given $\varepsilon > 0$ we can choose a compact neighborhood S_ε of $S_{\omega_2} \cap \mathcal{N}$ in S_2 such that if $X \in S_\varepsilon$, then $|\lambda| \leq \varepsilon$ for all eigenvalues λ of $\mathrm{ad}(X)$. On the other hand, we can choose $\beta > 0$ such that $|\lambda| \geq \beta$ for every nonzero eigenvalue λ of $\mathrm{ad}(H_2)$ for every $H_2 \in \omega_2$. Then if

$$\phi(x^*, H_2) = \varpi^{-\nu(x^*.H_2)} \cdot x^*.H_2 \in S_\varepsilon$$

for $x^* \in G^* = G/A_2$ and $H_2 \in \omega_2$, it is clear that

$$|x^*.H_2| \geq \beta/\varepsilon > q^{N-\mathfrak{f}_0}$$

if ε is sufficiently small.

Put $S' = S_{\omega_2} \cap {}^c S_\varepsilon$. Then S' is a compact set. We claim that there exists a compact set C^* in G^* such that

$$\phi({}^c C^* \times \omega_2) \subset S_\varepsilon.$$

If this were not the case, we could choose sequences $\{x_n^*\}$ in G^* and $\{H_n'\}$ in ω_2 such that $x_n^* \to \infty$ and $X_n = \phi(x_n^*, H_n') \in S'$. Since S' and ω_2 are compact, we may assume that $X_n \to X_0$ and $H_n' \to H_0$. Then X_0 is not nilpotent. Hence $|x_n^* \cdot H_n|$ remains bounded, and so, by standard results, x_n^* remains bounded contradicting our hypothesis that $x_n^* \to \infty$.

Now suppose $x^* \in {}^c C^*$, $H_1 \in \omega_1$, and $H_2 \in \omega_2$. Then

$$\int_K \chi\big(B(k.H_1, x^*.H_2)\big)\, dk = \int_K \chi\big(\varpi^{\nu(x^*.H_2)} \cdot B(k.H_1, X)\big)\, dk$$

where $X = \varpi^{-\nu(x^*.H_2)} \cdot x^*.H_2 = \phi(X^*, H_2) \in S_\varepsilon \subset S_2$. Hence $|x^*.H_2| > q^{N-\mathfrak{f}_0}$, and therefore

$$\nu(x^*.H_2) + N - \mathfrak{f}_0 < 0.$$

But then $N + \nu(x^*.H_2) < \mathfrak{f}_0$, and so

$$\int_K \chi\big(B(k.H_1, x^*.H_2)\big)\, dk = 0$$

by previous work. This completes the proof of our lemma. $\square$

Put

$$\Phi(H_1, H_2) = |\eta(H_1)|^{1/2} \cdot |\eta(H_2)|^{1/2} \int_{G/A_2} dx^* \int_K \chi\big(B(k.H_1, x.H_2)\big)\, dk$$

for $H_i \in \mathfrak{h}_i{}'$.

COROLLARY 7.6.   *The function Φ is locally constant on $\mathfrak{h}_1' \times \mathfrak{h}_2'$.*

PROOF.   This is immediate from Lemma 7.1. $\square$

THEOREM 7.7 (Theorem 13).   *The function Φ is locally bounded on $\mathfrak{h}_1 \times \mathfrak{h}_2$.*

REMARK 7.8.   If we prove the following lemma, then Theorem 7.7 would imply Theorem 6.1 (which is true if and only if Theorem 6.5 is true).

LEMMA 7.9 (Lemma 19).   *For $H_i \in \mathfrak{h}_i{}'$ we have*

$$\Phi(H_1, H_2) = |\eta(H_1)|^{1/2} \cdot |\eta(H_2)|^{1/2} \cdot \widehat{\mu_{H_2}}(H_1).$$

PROOF.   For $H_2 \in \mathfrak{h}_2'$ and $f \in \mathcal{D}$, put $T_{H_2}(f) = \phi_f^{\mathfrak{h}_2}(H_2)$. It follows from Theorem 1.1 that T_{H_2} is a locally summable function on $\mathfrak{g}$. Consider the mapping $(x, H) \mapsto x.H$ of $G \times \mathfrak{h}_1'$ into $\mathfrak{g}$. This map is everywhere submersive. Hence we have a linear, surjective mapping

$$\alpha \mapsto f_\alpha$$

of $C_c^\infty(G \times \mathfrak{h}_1')$ onto $C_c^\infty(\mathfrak{g}_{\mathfrak{h}_1})$ such that

$$\int_{\mathfrak{g}} f_\alpha(X) \cdot F(X)\, dX = \int_{G \times \mathfrak{h}_1'} \alpha(x, H) \cdot F(x.H)\, dx\, dH$$

for all $F \in C(\mathfrak{g})$. Fix $\beta \in C_c^\infty(\mathfrak{h}_1')$ and define $\alpha(x : H) = \delta(x) \cdot \beta(H)$ where δ is the characteristic function of K. Then

$$\phi_{f_\alpha}^{\mathfrak{h}_2}(H_2) = |\eta(H_2)|^{1/2} \int_{G/A_2} \widehat{f_\alpha}(x.H_2)\, dx^*$$

$$= |\eta(H_2)|^{1/2} \int_{G/A_2} dx^* \int_{\mathfrak{g}} \chi\big(B(X, x.H_2)\big) \cdot f_\alpha(X)\, dX$$

$$= |\eta(H_2)|^{1/2} \int_{G/A_2} dx^* \int_{K \times \mathfrak{h}_1} \chi\big(B(k.H_1, x.H_2)\big) \cdot \beta(H_1)\, dk\, dH_1$$

$$= \int_{\mathfrak{h}_1} |\eta(H_1)|^{-1/2} \cdot \Phi(H_1 : H_2) \cdot \beta(H_1)\, dH_1.$$

On the other hand,

$$\phi_{f_\alpha}^{\mathfrak{h}_2}(H_2) = T_{H_2}(f_\alpha) = \int_{\mathfrak{g}} T_{H_2}(X) \cdot f_\alpha(X)\, dX$$

$$= \int_{G \times \mathfrak{h}_1} \alpha(x : H_1) \cdot T_{H_2}(x.H_1)\, dx\, dH_1$$

$$= \int_{K \times \mathfrak{h}_1} \beta(H_1) \cdot T_{H_2}(H_1)\, dk\, dH_1$$

$$= \int_{\mathfrak{h}_1} \beta(H_1) \cdot T_{H_2}(H_1)\, dH_1.$$

Since this holds for all $\beta \in C_c^\infty(\mathfrak{h}_1')$, we must have

$$T_{H_2}(H_1) = |\eta(H_1)|^{-1/2} \cdot \Phi(H_1 : H_2)$$

or

$$\Phi(H_1 : H_2) = |\eta(H_1)|^{1/2} \cdot |\eta(H_2)|^{1/2} \cdot \widehat{\mu_{H_2}}(H_1). \quad \square$$

We shall now prove Theorems 6.1, 6.5, and 7.7 by induction on $\dim \mathfrak{g}$. By Lemma 6.6 and Remark 7.8, it is enough to prove Theorem 7.7. Let $\mathfrak{g}_1$ be the derived algebra of $\mathfrak{g}$. For $H_i \in \mathfrak{h}_i'$ and $Z_i \in \mathfrak{z}$ we have

$$|\eta(H_i + Z_i)| = |\eta(H_i)|,$$

and therefore

$$\Phi(H_1 + Z_1 : H_2 + Z_2)$$
$$= |\eta(H_1)|^{1/2} \cdot |\eta(H_2)|^{1/2} \int_{G/A_2} dx^* \int_K \chi\big(B(k.(H_1 + Z_1), x.(H_2 + Z_2))\big)\, dk$$
$$= \chi\big(B(Z_1, Z_2)\big) \cdot \Phi(H_1 : H_2).$$

Hence it is sufficient to prove Theorem 7.7 in the case where $\mathfrak{g}$ is semisimple.

Fix $H_2 \in \mathfrak{h}_2'$ and $\gamma \neq 0$ in $\mathfrak{h}_1$. Then it follows from Corollary 6.13 that $|\eta(H_1)|^{1/2} \cdot T_{H_2}(H_1) = \Phi(H_1 : H_2)$ remains bounded for $H_1 \in \mathfrak{h}_1'$ near γ. Hence we get the following result.

LEMMA 7.10. *Fix $H_2 \in \mathfrak{h}_2'$ and let ω_1 be a compact subset of $\mathfrak{h}_1$ such that $0 \notin \omega_1$. Put $\omega_1' = \omega_1 \cap \mathfrak{h}_1'$. Then*

$$\sup_{H_1 \in \omega_1'} |\Phi(H_1 : H_2)| < \infty.$$

PROOF. This is clear from the discussion preceding the lemma. $\square$

LEMMA 7.11. *Fix compact sets $\omega_1 \subset \mathfrak{h}_1$ and $\omega_2 \subset \mathfrak{h}_2$. Let $\omega_i' = \omega_i \cap \mathfrak{h}_i'$. Then we can choose $c > 0$ and a finite number of elements $\gamma_1, \gamma_2, \ldots, \gamma_r \in \omega_2'$ such that*

$$|\Phi(H_1 : H_2)| \leq c \sum_{i=1}^{r} |\Phi(H_1 : \gamma_i)|$$

for $H_i \in \omega_i'$.

PROOF. Let U be an open compact subset of $\mathfrak{g}$ containing ω_1, and let L be a lattice in $\mathfrak{g}$ such that $\hat{f} \in C_c(\mathfrak{g}/L)$ whenever $f \in C_c^\infty(U)$. For $H \in \omega_2'$, let σ_H denote the distribution on U given by

$$\sigma_H(f) = \phi_{\hat{f}}^{\mathfrak{h}_2}(H) = \phi_{\hat{f}}(H) = |\eta(H)|^{1/2} \cdot \mu_H(\hat{f})$$

for $f \in C_c^\infty(U)$. By Theorem 12.1, we know that the set of all σ_H $(H \in \omega_2')$ span a finite dimensional space (Lemma 1.2). We can choose $\gamma_1, \gamma_2, \ldots, \gamma_r \in \omega_2'$ such that $\sigma_i = \sigma_{\gamma_i}$ form a base for this space. Choose $f_j \in C_c^\infty(U)$ such that $\sigma_i(f_j) = \delta_{ij}$. Then, for $H \in \omega_i$,

$$\sigma_H = \sum_{i=1}^r \sigma_H(f_i) \cdot \sigma_i.$$

Since $\omega_1 \subset U$, we conclude from Lemma 7.9 that

$$\begin{aligned}
\Phi(H_1 : H_2) &= |\eta(H_1)|^{1/2} \cdot |\eta(H_2)|^{1/2} \cdot \widehat{\mu_{H_2}}(H_1) \\
&= |\eta(H_1)|^{1/2} \cdot \sigma_{H_2}(H_1) \\
&= \sum_{i=1}^r |\eta(H_1)|^{1/2} \cdot \sigma_{H_2}(f_i) \cdot \sigma_i(H_1) \\
&= \sum_{i=1}^r \phi_{\hat{f}_i}^{\mathfrak{h}_2}(H_2) \cdot \Phi(H_1 : \gamma_i)
\end{aligned}$$

for $H_i \in \omega_i'$. But we know [21, Theorem 13] that

$$\sup_{H_2 \in \omega_2'} \left| \phi_{\hat{f}}(H_2) \right| < \infty$$

for $f \in \mathcal{D}$. Therefore, there exists a $c > 0$ such that

$$|\Phi(H_1 : H_2)| \le c \sum_i |\Phi(H_1 : \gamma_i)|$$

for all $H_i \in \omega_i'$. $\qquad\square$

COROLLARY 7.12 (Lemma 20). *Suppose* $0 \notin \omega_1$. *Then* Φ *remains bounded on* $\omega_1' \times \omega_2'$

PROOF. This is immediate if we take into account Lemma 7.10. $\qquad\square$

In order to prove Theorem 7.7, we have to verify that Φ remains bounded on $\omega_1' \times \omega_2'$ (in the notation of Lemma 7.11). Without loss of generality, we may assume that ω_i is the set of all $H \in \mathfrak{h}_i$ such that $|H| \le T$ where $T \in \mathbb{R}$ and $T \ge 1$. Let S_1 be the set of all $H_1 \in \omega_1$ with $|H_1| \ge 1$. Then S_1 is a compact subset of ω_1 and $0 \notin S_1$. Hence, by Corollary 7.12, there exists a $c \in \mathbb{R}$ such that $|\Phi| \le c$ on $S_1' \times \omega_2'$ ($S_1' = S_1 \cap \mathfrak{h}_1'$). Now every element in ω_1' is of the form tH_1 where $H_1 \in S_1'$, $t \in \Omega^\times$, and $|t| \le 1$. Moreover $t\omega_2' \subset \omega_2'$, and

$$\Phi(tH_1 : H_2) = \Phi(H_1 : tH_2).$$

Hence $|\Phi(tH_1 : H_2)| \le c$ for all $H_2 \in \omega_2'$. Consequently, $|\Phi| \le c$ on $\omega_1' \times \omega_2'$. $\qquad\square$

Theorems 7.7, 6.5, 6.1, 4.4 and 1.1 are now fully established.

8. Some results on Shalika's germs

Fix a Cartan subalgebra $\mathfrak{h}$ of $\mathfrak{g}$. The following theorem is a slightly sharpened form of a result of Shalika [**66**].

THEOREM 8.1 (Theorem 14). *There exist functions* $\Gamma_{\mathcal{O}}$, *indexed by* $\mathcal{O} \in \mathcal{O}(0)$, *on* $\mathfrak{h}'$ *with the following properties:*

(1) $\Gamma_{\mathcal{O}}(t^2 H) = |t|^{r(\mathcal{O})-\ell} \Gamma_{\mathcal{O}}(H)$ *for* $t \in \Omega^{\times}$ *and* $H \in \mathfrak{h}'$.

(2) *For every* $f \in \mathcal{D}$, *we can choose a neighborhood* ω *of* 0 *in* $\mathfrak{h}$ *such that*

$$\phi_f^{\mathfrak{h}} = \sum_{\mathcal{O}} \mu_{\mathcal{O}}(f) \cdot \Gamma_{\mathcal{O}}$$

on $\omega \cap \mathfrak{h}'$.

These functions are unique. They take only real values and they satisfy, in addition, the following conditions.

(a) $\Gamma_{\mathcal{O}}(Z + H) = \Gamma_{\mathcal{O}}(H)$ *for* $Z \in \mathfrak{z}$ *and* $H \in \mathfrak{h}'$.

(b) $\Gamma_{\mathcal{O}}$ *is locally constant on* $\mathfrak{h}'$.

(c) $\Gamma_{\mathcal{O}}$ *is locally bounded on* $\mathfrak{h}$.

We shall call $\Gamma_{\mathcal{O}}$ the $\mathcal{O}$-germ on $\mathfrak{h}$. The proof of Theorem 8.1 is broken down into parts.

LEMMA 8.2. *Let* $\mathfrak{h}$ *be a Cartan subalgebra and* L *a lattice in* $\mathfrak{g}$. *Then there exists an open neighborhood* ω *of* 0 *in* $\mathfrak{h}$ *and functions* $\Gamma_{\mathcal{O}}$ *on* $\omega' = \omega \cap \mathfrak{h}'$ *such that*

$$\phi_f = \sum_{\mathcal{O} \in \mathcal{O}(0)} \mu_{\mathcal{O}}(f) \cdot \Gamma_{\mathcal{O}}$$

on ω' *for all* $f \in C_c(\mathfrak{g}/L)$.

PROOF. Fix a compact open neighborhood ω_0 of 0 in $\mathfrak{h}$. Then by Lemma 5.8 we can choose a lattice L_0 in $\mathfrak{g}$ with the following property. If $T \in J(\omega_0)$, then we can find complex numbers $c_{\mathcal{O}}(T)$ such that

$$j_{L_0} T = \sum_{\mathcal{O}} c_{\mathcal{O}}(T) \cdot j_{L_0} \mu_{\mathcal{O}}.$$

For $H \in \omega_0' = \omega_0 \cap \mathfrak{h}'$ and $f \in \mathcal{D}$, put

$$\tau_H(f) = \phi_f(H).$$

Then $\tau_H \in J(\omega_0)$. Put

$$\gamma_{\mathcal{O}}(H) = c_{\mathcal{O}}(\tau_H).$$

Then, for $f \in C_c(\mathfrak{g}/L_0)$ and $H \in \omega_0'$,

$$\phi_f(H) = \sum_{\mathcal{O}} \mu_{\mathcal{O}}(f) \cdot \gamma_{\mathcal{O}}(H).$$

Now choose $t \in \Omega^{\times}$ such that $L_0 \subset t^2 L$. Then, if $f \in C_c(\mathfrak{g}/L)$, we have

$$f_{t^2} \in C_c(\mathfrak{g}/t^2 L) \subset C_c(\mathfrak{g}/L_0)$$

and therefore

$$\phi_{f_{t^2}}(H) = \sum_{\mathcal{O}} \mu_{\mathcal{O}}(f_{t^2}) \cdot \gamma_{\mathcal{O}}(H) = \sum_{\mathcal{O}} \mu_{\mathcal{O}}(f) \cdot |t|^{d(\mathcal{O})} \cdot \gamma_{\mathcal{O}}(H)$$

for $H \in \omega_0'$. But

$$\phi_{f_{t^2}}(H) = |t|^{n-\ell} \phi_f(t^{-2}H).$$

Therefore

$$\phi_f(t^{-2}H) = \sum_{\mathcal{O}} \mu_{\mathcal{O}}(f) \cdot |t|^{d+\ell-n} \cdot \gamma_{\mathcal{O}}(H)$$

for $H \in \omega_0'$ and $f \in C_c(\mathfrak{g}/L)$. Now put $\omega = t^{-2}\omega_0$. Define, for $H \in \omega'$,

$$\Gamma_{\mathcal{O}}(H) = |t|^{d+\ell-n} \gamma_{\mathcal{O}}(t^2 H).$$

Then

$$\phi_f(H) = \sum_{\mathcal{O}} \mu_{\mathcal{O}}(f) \cdot \Gamma_{\mathcal{O}}(H)$$

for $H \in \omega'$ and $f \in C_c(\mathfrak{g}/L)$. $\qquad\square$

COROLLARY 8.3. *Notation as above. $\Gamma_{\mathcal{O}}$ is locally constant on ω' and bounded on ω.*

PROOF. If $\mathcal{O}_1$, $\mathcal{O}_2$, ..., $\mathcal{O}_r$ are the distinct nilpotent orbits in $\mathcal{O}(0)$ and we have $d(\mathcal{O}_i) = d_i$, then by Lemma 5.9 there exist real-valued functions $g_j \in C_c(\mathfrak{g}/L)$ such that

$$\mu_i(g_j) = \delta_{ij}.$$

If $\Gamma_i = \Gamma_{\mathcal{O}_i}$, then

$$\phi_{g_j}(H) = \sum_{i=1}^{r} \mu_i(g_j) \cdot \Gamma_i(H) = \Gamma_j(H).$$

This implies our assertion. $\qquad\square$

COROLLARY 8.4. *Fix $t \in \Omega^\times$ such that $t^2 L \subset L$ and $t^2 \omega \subset \omega$ (notation of Lemma 8.2). Then $\Gamma_{\mathcal{O}}(t^2 H) = |t|^{r(\mathcal{O})-\ell} \Gamma_{\mathcal{O}}(H)$ for $H \in \omega'$.*

PROOF. Fix $f \in C_c(\mathfrak{g}/L)$. Since $t^{-2}L \supset L$,

$$f_{t^{-2}} \in C_c(\mathfrak{g}/t^{-2}L) \subset C_c(\mathfrak{g}/L).$$

Hence

$$\phi_{f_{t^{-2}}}(H) = \sum_{\mathcal{O}} \mu_{\mathcal{O}}(f_{t^{-2}}) \cdot \Gamma_{\mathcal{O}}(H) = \sum_{\mathcal{O}} |t|^{-d(\mathcal{O})} \mu_{\mathcal{O}}(f) \cdot \Gamma_{\mathcal{O}}(H)$$

for $H \in \omega'$. But

$$\phi_{f_{t^{-2}}}(H) = |t|^{\ell-n} \phi_f(t^2 H) = |t|^{\ell-n} \sum_{\mathcal{O}} \mu_{\mathcal{O}}(f) \cdot \Gamma_{\mathcal{O}}(t^2 H).$$

Since the $\mu_{\mathcal{O},L}$ are linearly independent (Lemma 5.9), we conclude that for all $H \in \omega'$ we have $\Gamma_{\mathcal{O}}(t^2 H) = |t|^{n-\ell-d(\mathcal{O})} \Gamma_{\mathcal{O}}(H)$. $\qquad\square$

PROOF OF THEOREM 8.1. Fix a lattice L in $\mathfrak{g}$ such that $t^2 L \subset L$ for $t \in R$. Choose an open neighborhood ω of 0 in $\mathfrak{h}$ so that the conditions of Lemma 8.2 hold. Without loss of generality we may assume that $t^2 \omega \subset \omega$ for $t \in R$. Let $\mathcal{O}_1, \mathcal{O}_2, \ldots, \mathcal{O}_r$ be all the distinct nilpotent orbits. Put $\mu_i = \mu_{\mathcal{O}_i}$. Choose real-valued $f_i \in C_c(\mathfrak{g}/L)$ such that $\mu_i(f_j) = \delta_{ij}$ (Lemma 5.9). Put

$$\Gamma_i(H) = \phi_{f_i}(H)$$

for $H \in \omega' = \omega \cap \mathfrak{h}'$. Then we have seen that

$$\phi_f = \sum_i \mu_i(f) \cdot \Gamma_i$$

on ω' for $f \in C_c(\mathfrak{g}/L)$ and

$$\Gamma_i(t^2 H) = |t|^{r_i - \ell} \Gamma_i(H)$$

for all $t \in R$ and $H \in \omega'$. Hence Γ_i can be extended uniquely on $\mathfrak{h}'$ so that

$$\Gamma_i(t^2 H) = |t|^{r_i - \ell} \Gamma_i(H)$$

for all $t \in \Omega^\times$ and $H \in \mathfrak{h}'$.

Given $f \in \mathcal{D}$, we can choose $t \in R'$ such that $f \in C_c(\mathfrak{g}/t^2 L)$. Then $f_{t^{-2}} \in C_c(\mathfrak{g}/L)$, hence

$$\phi_{f_{t^{-2}}} = \sum_i \mu_i(f_{t^{-2}}) \cdot \Gamma_i$$

on ω'. From this we deduce immediately that $\phi_f = \sum_i \mu_i(f) \cdot \Gamma_i$ on $t^2 \omega'$. This proves existence.

Uniqueness follows from (1) and the fact that $\phi_{f_i} = \Gamma_i$ sufficiently near 0.

Corollary 8.3 shows that $\Gamma_i = \phi_{f_i}$ on ω'. Hence Γ_i is locally constant on ω' and therefore by (1) also on $\mathfrak{h}'$. In the same way we see that Γ_i is locally bounded on $\mathfrak{h}$ and $\Gamma_{\mathcal{O}}(Z + H) = \Gamma_{\mathcal{O}}(H)$ for all $Z \in \mathfrak{z}$ and $H \in \mathfrak{h}'$. $\qquad\square$

Let $A = A_{\mathfrak{h}}$ denote the split component of the Cartan subgroup of G corresponding to $\mathfrak{h}$. Let M and $\mathfrak{m}$ be the centralizers of A in G and $\mathfrak{g}$, respectively. Let $\eta_{\mathfrak{m}}$ be the polynomial function on $\mathfrak{m}$ which is the analogue of η on $\mathfrak{g}$. Then $\mathfrak{h}' \subset \mathfrak{h}''$ where $\mathfrak{h}''$ is the set of all points $H \in \mathfrak{h}$ with $\eta_{\mathfrak{m}}(H) \neq 0$.

For any nilpotent M-orbit ξ in $\mathfrak{m}$, let ν_ξ denote the corresponding M-invariant measure on $\mathfrak{m}$ and Γ_ξ the function on $\mathfrak{h}''$ defined by Theorem 8.1 (applied to $(M, \mathfrak{h})$ instead of $(G, \mathfrak{h})$).

LEMMA 8.5. *We have $\phi_f^{\mathfrak{h}} = \sum_\xi \nu_\xi(f_P) \cdot \Gamma_\xi$ near 0. Here ξ runs over all nilpotent M-orbits in $\mathfrak{m}$.*

PROOF. Since $\phi_f^{\mathfrak{h}} = \phi_{f_P}^{M/\mathfrak{h}}$, this is obvious. $\qquad\square$

COROLLARY 8.6. *The function $\Gamma_{\mathcal{O}}$ is a linear combination of the Γ_ξ where ξ runs over those nilpotent M-orbits in $\mathfrak{m}$ for which $r(\xi) = r(\mathcal{O})$.*

PROOF. We have seen that we can choose $f \in \mathcal{D}$ such that $\Gamma_{\mathcal{O}} = \phi_f^{\mathfrak{h}}$ near 0. Hence

(8.1) $$\Gamma_{\mathcal{O}} = \phi_f^{\mathfrak{h}} = \phi_{f_P}^{M/\mathfrak{h}} = \sum_{\xi} \nu_{\xi}(f_P) \cdot \Gamma_{\xi}$$

near 0. By the homogeneity condition on $\Gamma_{\mathcal{O}}$ and Γ_{ξ} (Theorem 8.1), only those ξ such that $r(\xi) = r(\mathcal{O})$ can occur on the right side of equation (8.1). Since equation (8.1) holds near 0 and the Γ_i's are unique, it must hold on $\mathfrak{h}'$. $\square$

COROLLARY 8.7. *The function $\Gamma_{\mathcal{O}}$ is locally constant on $\mathfrak{h}''$.*

PROOF. Clear. $\square$

COROLLARY 8.8. *The function $\Gamma_{\mathcal{O}} = 0$ if $d(\mathcal{O}) < \dim(G/M)$.*

PROOF. Let $n = \dim(G)$ and $m = \dim(M)$. Then $r(\mathcal{O}) = n - d(\mathcal{O})$ and $r(\xi) = m - d(\xi)$. Therefore $\Gamma_{\mathcal{O}} = 0$ unless $r(\mathcal{O}) = r(\xi)$ for some ξ (Corollary 8.6). This implies that

$$d(\mathcal{O}) = n - m + d(\xi) \geq n - m. \quad \square$$

Let 0 denote the orbit $\{0\}$.

COROLLARY 8.9. *The function $\Gamma_0 = 0$ unless $\mathfrak{h}$ is elliptic.*

PROOF. $d(0) = 0$ and so $\Gamma_0 = 0$ unless $M = G$ (i.e., unless $\mathfrak{h}$ is elliptic). $\square$

9. Proof of Theorem 9.6

For $\mathcal{O} \in \mathcal{O}(0)$ put

$$\Psi_{\mathcal{O}} = |\eta|^{1/2} \cdot \widehat{\mu_{\mathcal{O}}}.$$

For any Cartan subalgebra $\mathfrak{h}$, let $\Gamma_{\mathcal{O}}^{\mathfrak{h}}$ denote the $\mathcal{O}$-germ on $\mathfrak{h}$ as in Theorem 8.1. We return to the notation of Theorem 7.7.

THEOREM 9.1 (Lemma 21). *Fix a compact subset ω_1 of $\mathfrak{h}_1$. Then we can choose a neighborhood ω_2 of 0 in $\mathfrak{h}_2$ such that*

$$\Phi(H_1 : H_2) = \sum_{\mathcal{O} \in \mathcal{O}(0)} \Psi_{\mathcal{O}}(H_1) \cdot \Gamma_{\mathcal{O}}^{\mathfrak{h}_2}(H_2)$$

for $H_i \in \omega_i' = \omega_i \cap \mathfrak{h}_i'$.

PROOF. We may assume that ω_1 is open in $\mathfrak{h}_1$. Let U be a compact open subset of $\mathfrak{g}$ containing ω_1. Fix a lattice L in $\mathfrak{g}$ such that $\hat{f} \in C_c(\mathfrak{g}/L)$ for $f \in C_c^{\infty}(U)$. Then, by Lemma 8.2, we can choose a neighborhood ω_2 of 0 in $\mathfrak{h}_2$ such that, for $h \in C_c(\mathfrak{g}/L)$,

$$\phi_h^{\mathfrak{h}_2} = \sum_{\mathcal{O}} \mu_{\mathcal{O}}(h) \cdot \Gamma_{\mathcal{O}}^{\mathfrak{h}_2}$$

on ω_2'. Hence

$$\phi_{\hat{f}}^{\mathfrak{h}_2}(H_2) = \sum_{\mathcal{O}} \widehat{\mu_{\mathcal{O}}}(f) \cdot \Gamma_{\mathcal{O}}^{\mathfrak{h}_2}(H_2)$$

for $f \in C_c^\infty(U)$ and $H_2 \in \omega_2'$.

Here we follow the proof of Lemma 7.9. Fix a compact open subgroup K of G such that $\omega_1^K \subset U$. The mapping $(x, H) \mapsto x.H$ of $G \times \mathfrak{h}_1'$ into $\mathfrak{g}$ is everywhere submersive. Hence, there exists a linear mapping $\alpha \mapsto f_\alpha$ of $C_c^\infty(G \times \mathfrak{h}_1')$ into $\mathcal{D}$ such that

$$\int_{G \times \mathfrak{h}_1'} \alpha(x : H) \cdot F(x.H) \, dx \, dH = \int_{\mathfrak{g}} f_\alpha(X) \cdot F(X) \, dX$$

for $F \in C(\mathfrak{g})$. (We normalize the Haar measure on G in such a way that $\int_K dx = 1$.) Fix $\beta \in C_c^\infty(\omega_1')$ and define $\alpha(x : H) = \delta(x) \cdot \beta(H)$ where δ is the characteristic function of K. Then

$$\operatorname{Supp}(f_\alpha) \subset \omega_1^K \subset U.$$

Put $f = f_\alpha$. Then

$$\phi_f^{\mathfrak{h}_2}(H_2) = \sum_{\mathcal{O}} \widehat{\mu_{\mathcal{O}}}(f) \cdot \Gamma_{\mathcal{O}}^{\mathfrak{h}_2}(H_2)$$

for $H_2 \in \omega_2'$. But

$$\widehat{\mu_{\mathcal{O}}}(f) = \int_{\mathfrak{g}} \widehat{\mu_{\mathcal{O}}}(X) \cdot f(X) \, dX$$

$$= \int_K dk \int_{\mathfrak{h}_1} \beta(H_1) \cdot \widehat{\mu_{\mathcal{O}}}(H_1) \, dH_1$$

$$= \int_{\mathfrak{h}_1} \beta(H_1) \cdot \widehat{\mu_{\mathcal{O}}}(H_1) \, dH_1.$$

On the other hand, the proof of Lemma 7.9 shows that

$$\phi_{\hat f}^{\mathfrak{h}_2}(H_2) = \int_{\mathfrak{h}_1} \tau_{H_2}(H_1) \cdot \beta(H_1) \, dH_1$$

and

$$|\eta(H_1)|^{1/2} \cdot \tau_{H_2}(H_1) = \Phi(H_1 : H_2)$$

where $\tau_{H_2}(g) = \phi_{\hat g}^{\mathfrak{h}_2}(H_2)$ for $g \in \mathcal{D}$. Consequently, for $H_i \in \omega_i'$,

$$\Phi(H_1 : H_2) = |\eta(H_1)|^{1/2} \cdot \tau_{H_2}(H_1)$$

$$= \sum_{\mathcal{O}} |\eta(H_1)|^{1/2} \cdot \widehat{\mu_{\mathcal{O}}}(H_1) \cdot \Gamma_{\mathcal{O}}^{\mathfrak{h}_2}(H_2)$$

$$= \sum_{\mathcal{O}} \Psi_{\mathcal{O}}(H_1) \cdot \Gamma_{\mathcal{O}}^{\mathfrak{h}_2}(H_2). \quad \square$$

LEMMA 9.2 (Lemma 22). *If $\mathcal{O}$ is any nilpotent G-orbit in $\mathfrak{g}$ and $\ell = \operatorname{rank}(\mathfrak{g})$, then*

$$\Psi_{\mathcal{O}}(t^2 X) = |t|^{r(\mathcal{O})-\ell} \Psi_{\mathcal{O}}(X)$$

for $t \in \Omega^\times$ and $X \in \mathfrak{g}'$.

PROOF. For $f \in \mathcal{D}$ and $t \in \Omega^\times$, we have $(f_{t^2})^\wedge = |t|^{2n} (\hat{f})_{t^{-2}}$ where $n = \dim(\mathfrak{g})$. Hence

$$\widehat{\mu_\mathcal{O}}(f_{t^2}) = \mu_\mathcal{O}\big((f_{t^2})^\wedge\big) = |t|^{2n} \mu_\mathcal{O}\big((\hat{f})_{t^{-2}}\big)$$
$$= |t|^{2n-d} \mu_\mathcal{O}(\hat{f}) = |t|^{2n-d} \widehat{\mu_\mathcal{O}}(f).$$

But

$$\widehat{\mu_\mathcal{O}}(f_{t^2}) = |t|^{2n} \int_{\mathfrak{g}} \widehat{\mu_\mathcal{O}}(t^2 X) \cdot f(X) \, dX.$$

Hence we conclude that

$$\widehat{\mu_\mathcal{O}}(t^2 X) = |t|^{-d(\mathcal{O})} \widehat{\mu_\mathcal{O}}(X)$$

for $t \in \Omega^\times$ and $X \in \mathfrak{g}'$. Therefore

$$\Psi_\mathcal{O}(t^2 X) = |t|^{r(\mathcal{O})-\ell} \Psi_\mathcal{O}(X)$$

for $t \in \Omega^\times$. $\qquad\square$

As before, let $\mathfrak{g}_e$ be the set of all regular elliptic elements of $\mathfrak{g}$. Fix a compact open subgroup K of G and let dk denote the normalized Haar measure on K. Put

$$\Phi(X_1 : X_2) = |\eta(X_1)|^{1/2} \cdot |\eta(X_2)|^{1/2} \int_{G/Z} dx^* \int_K \chi\big(B(k.X_1, x.X_2)\big) \, dk$$

for $X_i \in \mathfrak{g}_e$.

LEMMA 9.3 (Lemma 23). *We have $\Phi \in C^\infty(\mathfrak{g}_e \times \mathfrak{g}_e)$ and $\Phi(X_1 : X_2) = \Phi(X_2 : X_1)$ for $X_i \in \mathfrak{g}_e$.*

PROOF. From Lemma 7.1, $\Phi \in C^\infty(\mathfrak{g}_e \times \mathfrak{g}_e)$. It is obvious that

$$\Phi(X_1 : X_2) = |\eta(X_1)|^{1/2} \cdot |\eta(X_2)|^{1/2} \int_{G/Z} dx^* \int_{K \times K} \chi\big(B(k_1.X_1, xk_2.X_2)\big) \, dk_1 \, dk_2$$

and from this the second assertion follows from the G-invariance and symmetry of B. $\qquad\square$

COROLLARY 9.4. *Let $\mathfrak{h}_1$ and $\mathfrak{h}_2$ be two elliptic Cartan subalgebras of $\mathfrak{g}$. Suppose $d \geq 0$ is an integer. Then*

$$\sum_{d(\mathcal{O})=d} \Psi_\mathcal{O}(H_1) \cdot \Gamma_\mathcal{O}^{\mathfrak{h}_2}(H_2) = \sum_{d(\mathcal{O})=d} \Psi_\mathcal{O}(H_2) \cdot \Gamma_\mathcal{O}^{\mathfrak{h}_1}(H_1)$$

for $H_i \in \mathfrak{h}_i'$. Here $\mathcal{O}$ runs over all nilpotent G-orbits in $\mathfrak{g}$ with $d(\mathcal{O}) = d$.

PROOF. From Theorem 8.1 and Lemma 9.2 we know that, for $t \in \Omega^\times$, and $H_i \in \mathfrak{h}_i'$,

$$\Gamma_\mathcal{O}^{\mathfrak{h}_i}(t^2 H_i) = |t|^{r(\mathcal{O})-\ell} \Gamma_\mathcal{O}^{\mathfrak{h}_i}(H_i)$$

while, for $X \in \mathfrak{g}'$ and $t \in \Omega^\times$,

$$\Psi_\mathcal{O}(t^2 X) = |t|^{r(\mathcal{O})-\ell} \Psi_\mathcal{O}(X).$$

From Lemma 9.3 and Theorem 9.1 we have

$$\sum_{\mathcal{O}} \Psi_{\mathcal{O}}(H_1) \cdot \Gamma_{\mathcal{O}}^{\mathfrak{h}_2}(H_2) = \sum_{\mathcal{O}} \Psi_{\mathcal{O}}(H_2) \cdot \Gamma_{\mathcal{O}}^{\mathfrak{h}_1}(H_1).$$

The result is now immediate.   □

LEMMA 9.5 (Lemma 24). *Let* $\mathfrak{h}_1, \mathfrak{h}_2, \ldots, \mathfrak{h}_r$ *be a complete set of Cartan subalgebras of* $\mathfrak{g}$ *no two of which are conjugate under* G. *Suppose that we have constants* $c_{\mathcal{O}} \in \mathbb{C}$ *such that*

$$\sum_{\mathcal{O} \in \mathcal{O}(0)} c_{\mathcal{O}} \cdot \Gamma_{\mathcal{O}}^{\mathfrak{h}_i} = 0$$

for $1 \leq i \leq r$. *Then* $c_{\mathcal{O}} = 0$ *for all* $\mathcal{O} \in \mathcal{O}(0)$.

PROOF. Since the $\mu_{\mathcal{O}}$ are linearly independent (Lemma 3.3), we can choose $f \in \mathcal{D}$ such that $\mu_{\mathcal{O}}(f) = c_{\mathcal{O}}$ for all $\mathcal{O} \in \mathcal{O}(0)$. Moreover, by Theorem 8.1, there exist neighborhoods ω_i of 0 in $\mathfrak{h}_i$ such that

$$\phi_f^{\mathfrak{h}_i} = \sum_{\mathcal{O}} \mu_{\mathcal{O}}(f) \cdot \Gamma_{\mathcal{O}}^{\mathfrak{h}_i} = 0$$

on $\omega_i' = \omega_i \cap \mathfrak{h}_i'$. By Lemma 2.1, we can choose a G-domain D in $\mathfrak{g}$ such that

$$0 \in D \cap \mathfrak{h}_i \subset \omega_i.$$

Let F denote the characteristic function of D. Put $f_0 = F \cdot f \in \mathcal{D}$. If $1 \leq i \leq r$ and $H_i \in \mathfrak{h}_i'$, then

$$\phi_{f_0}^{\mathfrak{h}_i}(H_i) = F(H_i) \cdot \phi_f^{\mathfrak{h}_i}(H_i) = 0.$$

Hence we conclude from Theorem 3.1 that $\mu_{\mathcal{O}}(f_0) = 0$ for $\mathcal{O} \in \mathcal{O}(0)$. But since $\mathrm{Supp}(\mu_{\mathcal{O}}) \subset \mathcal{N}$ and $F = 1$ on $\mathcal{N}$, we conclude that

$$c_{\mathcal{O}} = \mu_{\mathcal{O}}(f) = \mu_{\mathcal{O}}(f_0) = 0. \quad \square$$

LEMMA 9.6 (Theorem 15). *For every elliptic Cartan subalgebra* $\mathfrak{h}$ *of* $\mathfrak{g}$ *there exists a real number* $c \neq 0$ *such that*

$$\Gamma_0^{\mathfrak{h}}(H) = c \, |\eta(H)|^{1/2}$$

for $H \in \mathfrak{h}'$.

PROOF. Clearly $\mu_0(f) = f(0)$ for $f \in \mathcal{D}$ and therefore $\widehat{\mu_0} = 1$ and so $\Psi_0 = |\eta|^{1/2}$.

Now $\Gamma_0^{\mathfrak{h}} = 0$ if $\mathfrak{h}$ is a Cartan subalgebra of $\mathfrak{g}$ which is not elliptic (Corollary 8.9). It follows from Lemma 9.5 that we can choose an elliptic Cartan subalgebra $\mathfrak{h}_1$ such that $\Gamma_0^{\mathfrak{h}_1} \neq 0$.

Let $\mathfrak{h}_2$ be any elliptic Cartan subalgebra. Since 0 is the only orbit of dimension zero, we conclude from Corollary 9.4 that

$$|\eta(H_1)|^{-1/2} \cdot \Gamma_0^{\mathfrak{h}_1}(H_1) = |\eta(H_2)|^{-1/2} \cdot \Gamma_0^{\mathfrak{h}_2}(H_2)$$

for $H_i \in \mathfrak{h}_i'$. This shows that there exists a constant $c \in \mathbb{C}$ such that $\Gamma_0^{\mathfrak{h}_1}(H_1) = c |\eta(H_1)|^{1/2}$ for all $H_1 \in \mathfrak{h}_1'$. Since $\Gamma_0^{\mathfrak{h}_1}$ is real-valued and non-zero, it follows that c is real and non-zero.   □

Part II. An extension and proof of Howe's Theorem

10. Some special subsets of $\mathfrak{g}$

We first introduce some special subsets of $\mathfrak{g}$ that will be used throughout the remainder of this paper.

10.1. Admissible G-domains. Let $\bar{\Omega}$ be the algebraic closure of Ω. We extend the absolute value to $\bar{\Omega}$ as follows. If $x \in \bar{\Omega}$, let L be any subfield of $\bar{\Omega}$ containing $\Omega(x)$ such that $[L : \Omega] < \infty$. We define $|x|$ by

$$|x|^{[L:\Omega]} = \big|N_{L/\Omega}(x)\big|.$$

For any $X \in M_n(\bar{\Omega})$, put $\mu(X) = \max_\lambda |\lambda|$ where λ runs over all eigenvalues of X.

Fix a real number $\varepsilon \in (0,1)$ such that

$$\frac{\varepsilon^k}{|k!|} \to 0 \text{ as } k \to \infty$$

and

$$\varepsilon \geq \frac{\varepsilon^k}{|k!|}$$

for $k \geq 2$.

LEMMA 10.1. *Fix $X \in M_n(\bar{\Omega})$ such that $\mu(X) \leq \varepsilon$. Then*

$$\lim_{k \to \infty} \left| \frac{X^k}{k!} \right| = 0.$$

So $\sum_{k \geq 0} \frac{X^k}{k!}$ converges. If we define

$$\exp(X) = \sum_{k \geq 0} \frac{X^k}{k!},$$

then $\mu\big(\exp(X) - 1\big) \leq \varepsilon$. Conversely, if $X \in M_n(\bar{\Omega})$ and $\mu(X - 1) \leq \varepsilon$, then

$$\lim_{k \to \infty} \left| \frac{(X-1)^k}{k} \right| = 0.$$

So $\sum_{k \geq 1}(-1)^{k-1}\frac{(X-1)^k}{k}$ converges and if we define

$$\log(X) = \sum_{k \geq 1}(-1)^{k-1}\frac{(X-1)^k}{k},$$

then $\mu\big(\log(X)\big) \leq \varepsilon$. *Moreover, if* $y \in GL_n(\bar{\Omega})$, *then*

$$\exp\big(\mathrm{Ad}(y)X\big) = \mathrm{Ad}(y)\exp(X)$$

and

$$\log\big(\mathrm{Ad}(y)X\big) = \mathrm{Ad}(y)\log(X).$$

PROOF. Let X_s and X_n be the semisimple and nilpotent components of X with respect to the Jordan decomposition. We can choose $y \in GL_n(\bar{\Omega})$ such that

$$\lambda = y^{-1}X_s y$$

is diagonal. Then

$$X^k = y(\lambda + Y)^k y^{-1}$$

where $Y = y^{-1}X_n y$ is nilpotent and λ and Y commute. Therefore, for $k \geq 0$,

$$|X|^k \leq c_1 \cdot \big|(\lambda + Y)^k\big| \leq c_1 \cdot \max_{0 \leq r < n} |\lambda|^{k-r} \cdot |Y^r|$$

$$\leq c_2 \cdot \varepsilon^k$$

where $c_2 = c_1 \cdot \varepsilon^{-n} \cdot \max_{0 \leq r < n} |Y^r|$. So

$$\left|\frac{X^k}{k!}\right| \leq c_2 \cdot \frac{\varepsilon^k}{|k!|} \to 0$$

as $k \to \infty$. Also

$$\left|\frac{\lambda^k}{k!}\right| \leq \frac{\varepsilon^k}{|k!|} \leq \varepsilon$$

for $k \geq 2$ and so $\mu\big(\exp(X) - 1\big) = \mu\big(\exp(\lambda) - 1\big) \leq \varepsilon$.

The statement for $\log(X)$ follows in the same way. The rest is clear. $\square$

COROLLARY 10.2. *If* $\mu(X) \leq \varepsilon$, *then* $\log\big(\exp(X)\big) = X$. *Similarly, if* $\mu(X - 1) \leq \varepsilon$, *then* $\exp\big(\log(X)\big) = X$.

PROOF. Note that $\exp(X) - 1 = \sum_{k \geq 1} \frac{X^k}{k!}$ and so

$$\big(\exp(X) - 1\big)^m = \sum_{k_1,\ldots,k_m \geq 1} \frac{X^{k_1 + \cdots + k_m}}{k_1! \ldots k_m!}.$$

As in the proof of Lemma 10.1, we can find $c \geq 1$ such that $|X^k| \leq c \cdot \varepsilon^k$ for $k \geq 0$. Then

$$\left|\frac{X^{k_1 + \cdots + k_m}}{k_1! \ldots k_m!}\right| \leq c \cdot \frac{\varepsilon^{k_1 + \cdots + k_m}}{|k_1! \ldots k_m!|} \leq c \cdot \varepsilon^m$$

for $k_i \geq 1$ $(1 \leq i \leq m)$. Since

$$\frac{\varepsilon^m}{|m|} \leq \frac{\varepsilon^m}{|m!|} \to 0$$

as $m \to \infty$, the double series

$$\sum_{m \geq 1} \frac{(-1)^{m-1}}{m} \sum_{k_1,\ldots,k_m \geq 1} \frac{X^{k_1 + \cdots + k_m}}{k_1! \ldots k_m!}$$

converges absolutely. So the result follows by formal rearrangement. The other case is handled similarly. $\square$

Let L be a lattice in $\mathfrak{gl}_n(\Omega)$. By taking L to be small, we may assume that $|X| \le \varepsilon$ and $\mu(X) \le \varepsilon$ for $X \in L$. Then the exponential mapping is defined on $L^{GL_n(\Omega)} = U$. Put $V = \exp(U)$. Then log is defined on V and both $\exp\colon U \to V$ and $\log\colon V \to U$ are bijective. exp is "analytic" on L and therefore by conjugacy also on U. Similarly for log on V.

Note that $\mu(\operatorname{ad} X) \le \mu(X)$. Hence the series

$$\sum_{k \ge 1} \left| \frac{(\operatorname{ad} X)^{k-1}}{k!} \right|$$

converges for $X \in U$. Define, for $X \in U$,

$$\xi(X) = \sum_{k \ge 1} \frac{(\operatorname{ad} X)^{k-1}}{k!}.$$

Now we return to our usual notation with $\mathfrak{g}$ and G. In particular, exp is a map from $\mathfrak{g}$ to G which is defined on a sufficiently small G-domain in $\mathfrak{g}$. A subset ω of $\mathfrak{g}$ will be called admissible if $\omega \subset \mathfrak{g} \cap U$ and

$$|\det \xi_{\mathfrak{g}}(X)| = 1$$

for $X \in \omega$. Here $\xi_{\mathfrak{g}}(X)$ is the restriction of $\xi(X)$ on $\mathfrak{g}$. If ω is admissible, it is clear that for $X \in \omega$

$$|\eta_{\mathfrak{g}}(X)| = \left| D_G\big(\exp(X)\big) \right|.$$

LEMMA 10.3. *We can choose a G-domain W in $\mathfrak{g}$ such that*

(1) *$0 \in W$ and W is admissible and*
(2) *for all compact subsets Q of G, $(\exp W) \cap QZ$ is compact.*

(Z is the maximal split torus lying in the center of G.)

PROOF. Let $\mathfrak{t}$ be the center and $\mathfrak{g}_1 = [\mathfrak{g}, \mathfrak{g}]$ the derived algebra of $\mathfrak{g}$. Let T be the maximal Ω-torus lying in the center of G and G_1 the derived group of G. Then $G_0 = G_1 T$ is an open subgroup of G, and therefore it is also closed.

Let ω be an open, compact, admissible neighborhood of 0 in $\mathfrak{t} \cap L + \mathfrak{g}_1 \cap L$. Choose a G-domain D_0 in $\mathfrak{g}$ such that $0 \in D_0 \subset \omega^G$. Put $D_1 = D_0 \cap \mathfrak{g}_1$, and choose a compact open neighborhood $\mathfrak{t}_0$ of 0 in $\mathfrak{t} \cap D_0$. Then $W = \mathfrak{t}_0 + D_1$ is obviously a G-domain in $\mathfrak{g}$ satisfying condition (1) of the lemma. Hence, $\exp(W)$ and $\exp(D_1)$ are closed in G, and $\exp(\mathfrak{t}_0)$ is compact.

Since T/Z is compact, it is enough to verify that $(\exp W) \cap QT$ is compact for any compact set Q in G. Since T is closed, QT is closed. Also

$$(\exp W) \cap QT = \big(\exp(D_1) \cap QT\big) \cdot \exp(\mathfrak{t}_0).$$

So it is enough to verify that $\exp(D_1) \cap QT$ is compact. Since $\exp(D_1) \subset G_1$ and $\exp(D_1)$ is closed, it would be enough to verify that $G_1 \cap QT$ is compact.

Let σ denote the projection of G on G/T. Then $\sigma(G_1) = \sigma(G_0)$ is an open subgroup of $\sigma(G)$ and so it is closed. Hence $\sigma(G_1 \cap QT) = \sigma(G_1) \cap \sigma(Q)$ is compact.

On the other hand, $G_1 \cap T$ is finite. Therefore $G_1 \cap QT$, being the complete inverse image of $\sigma(G_1) \cap \sigma(Q)$ in G_1, is compact. $\hfill\square$

DEFINITION 10.4. A G-domain $\mathfrak{g}_0 \subset \mathfrak{g}$ is said to be $(G, \mathfrak{g})$-admissible if

(1) $t \cdot \mathfrak{g}_0 \subset \mathfrak{g}_0$ for all $t \in R$,

(2) $\mathfrak{g}_0$ is admissible, and

(3) for all compact subsets Q of G, $\exp(\mathfrak{g}_0) \cap QZ$ is compact.

REMARK 10.5. It is clear from Lemma 10.3 that $(G, \mathfrak{g})$-admissible G-domains exist.

10.2. Adapted lattices. Fix a maximal split torus A in G. Let K be a good maximal compact open subgroup corresponding to A. Let $\chi(A)$ denote the group of all rational characters of A which are defined over Ω.

DEFINITION 10.6. A lattice L in $\mathfrak{g}$ is said to be adapted to (K, A) if there exists a base $X_1, X_2, \dots, X_n$ for $\mathfrak{g}$ over Ω and elements $\chi_i \in \chi(A)$ such that

(1) $\mathrm{Ad}(a)X_i = \chi_i(a)X_i$ for $a \in A$ and $1 \le i \le n$,

(2) $L = \sum_i RX_i$, and

(3) $L^k = L$ for all $k \in K$.

Let L^* be the lattice dual to L. Then L^* consists of all points $\mu \in \mathfrak{g}$ such that

$$\chi\big(B(\mu, \lambda)\big) = 1$$

for all $\lambda \in L$. It is easy to check that if L is adapted to (K, A), the same holds for L^*.

We shall say that L is well adapted if it is adapted to (K, A_0) for a suitable choice of (K, A_0).

REMARK 10.7. It is a consequence of a result of Bruhat-Tits that we can always produce a well adapted lattice. (If x is a special point in the Bruhat-Tits building of G and r is a real number, then the lattice $\mathfrak{g}_{x,r}$ defined in [**42, 43**] is well adapted.) Further, by multiplying this lattice by an element in R', it can be made arbitrarily small.

11. An extension of Howe's Theorem

11.1. The space $J(V, t, L)$. Define a norm on $\mathfrak{g}$ as in §2. Note that $|cX| = |c||X|$ and $|[X, Y]| \le |X||Y|$ for $c \in \Omega$ and $X, Y \in \mathfrak{g}$. Moreover, if $X \ne 0$, we can choose $c \in \Omega^\times$ such that $|cX| = 1$. For any non-empty subset ω of $\mathfrak{g}$, put

$$|\omega| = \sup_{X \in \omega} |X|.$$

For $t \ge 0$, let $\mathfrak{g}(t)$ denote the set of all $X \in \mathfrak{g}$ with $|X| \le t$. Moreover, let S be the set of all $X \in \mathfrak{g}$ with $|X| = 1$.

Let L be a lattice in $\mathfrak{g}$, V a neighborhood of $\mathcal{N} \cap S$ in S, and t a positive real number. For $X \in \mathfrak{g}$ let f_X be the characteristic function of the coset $X + L$ in $\mathfrak{g}$.

We denote by $J(V, t, L)$ the subspace of all $T \in J$ with the following property. If $X \in \mathfrak{g}$ and $|X| \geq t$, then $T(f_X) = 0$ unless $X \in \Omega V$.

Fix (V, t, L) as above. By $C(V, t, L)$ we mean the following condition on (V, t, L).

CONDITION $C(V, t, L)$.   *Suppose $X \in \Omega V$ and $|X| \geq t$. Put*

$$\phi_\alpha = \int_G \alpha(x) \cdot (f_X)^x \, dx$$

where f_X is the characteristic function of $X + L$ on $\mathfrak{g}$. Then we can choose $\alpha \in C_c^\infty(G)$ such that

(1) $\int_G \alpha(x) \, dx \neq 0$
(2) $\phi_\alpha \in C_c(\mathfrak{g}/L)$
(3) $\mathrm{Supp}(\phi_\alpha) \subset \mathfrak{g}(t')$ *for some t' such that $0 < t' < |X|$.*

The following theorem provides a justification for this condition.

THEOREM 11.1 (Theorem 16).   *If $C(V, t, L)$ holds, then $\dim j_L J(V, t, L) < \infty$.*

It is enough to verify the following lemma. The proof of it imitates Howe's original argument.

LEMMA 11.2 (Lemma 25).   *Suppose $C(V, t, L)$ holds and T is an element of $J(V, t, L)$ such that $T(f_X) = 0$ for all $X \in \mathfrak{g}(t)$. Then $j_L T = 0$.*

PROOF. Suppose $j_L T \neq 0$. Then $T(f_X) \neq 0$ for some $X \in \mathfrak{g}$. Clearly $|X| > t$. Choose X so that $T(f_X) \neq 0$ and $|X|$ has the least possible value. Then since $T \in J(V, t, L)$ and $|X| > t$, it follows that $X \in \Omega V$. Now choose $\alpha \in C_c^\infty(G)$ as in the statement of the condition $C(V, t, L)$. Then

$$T(\phi_\alpha) = \int_G \alpha(x) \, dx \cdot T(f_X) \neq 0.$$

On the other hand, $\phi_\alpha \in C_c(\mathfrak{g}/L)$ and $\mathrm{Supp}(\phi_\alpha) \subset \mathfrak{g}(t')$ with $0 < t' < |X|$. Therefore, it follows from the definition of X that $T(\phi_\alpha) = 0$. This contradiction proves the lemma.   □

It is obvious that

$$J(V, t, L) \subset J(V, t', L)$$

for $t' \geq t$. Put

$$J(V, \infty, L) = \bigcup_{t > 0} J(V, t, L).$$

If $t_0 \geq t$ and V_0 is a neighborhood of $\mathcal{N} \cap S$ in V, then it is clear that

$$C(V, t, L) \Rightarrow C(V_0, t_0, L).$$

Put

$$J_0 = \bigcup_\omega J(\omega)$$

where ω runs over all compact subsets of $\mathfrak{g}$. Let $J_0(V, t, L) = J_0 \cap J(V, t, L)$, and $J_0(V, \infty, L) = J_0 \cap J(V, \infty, L)$.

THEOREM 11.3 (Theorem 17). *Suppose $C(V, t, L)$ holds. Choose a lattice Λ in $\mathfrak{g}$ such that $|\Lambda| < 1$ and*

$$V_0 = (\mathcal{N} \cap S) + \Lambda \subset V.$$

Then

$$j_L J(V_0, t_0, L) = j_L J_0(V_0, t_0, L)$$

for all t_0 sufficiently large.

We shall give a proof in §11.2.

COROLLARY 11.4. *We have*

$$j_L J(V_0, \infty, L) = j_L J_0(V_0, \infty, L) \subset j_L J_0.$$

11.2. Proof of Theorem 11.3. Let $S(\mathfrak{g})$ be the symmetric algebra over $\mathfrak{g}$ and $I(\mathfrak{g})$ the subalgebra of all G-invariants in $S(\mathfrak{g})$. By means of the bilinear form B, we may identify elements of $S(\mathfrak{g})$ with polynomial functions on $\mathfrak{g}$.

LEMMA 11.5. *There exist nonzero homogeneous elements p_1, p_2, ..., p_r in $I(\mathfrak{g})$ with the following two properties.*

(1) *$\mathcal{N}$ is exactly the set of all points $X \in \mathfrak{g}$ where $p_i(X) = 0$ for $1 \leq i \leq r$.*
(2) *Let U be any subset of $\mathfrak{g}$ such that all the $|p_i|$ remain bounded on U. Then there exists a compact subset ω in $\mathfrak{g}$ such that $U \subset \omega^G$.*

PROOF. As usual, let $\mathfrak{z}$ denote the center of $\mathfrak{g}$. Fix a base $Z_1, \ldots, Z_s$ for $\mathfrak{z}$ over Ω, and let $\lambda_1, \lambda_2, \ldots, \lambda_s$ denote the linear functions

$$\lambda_i(X) = B(Z_i, X)$$

for $X \in \mathfrak{g}$. As on page 14 put

$$\det(t - \operatorname{ad} X) = \sum_{\ell \leq k \leq n} p'_k(X) t^k$$

for $X \in \mathfrak{g}$. Here t is an indeterminate. The p'_k are homogeneous elements in $I(\mathfrak{g})$. Let $p_1, p_2, \ldots, p_r$ denote all the distinct non-constant elements among the λ_i and p'_k. Then p_i is a homogeneous element in $I(\mathfrak{g})$. Note that condition (1) is satisfied.

We now contend with condition (2). Let d_i denote the degree of p_i. For $X \in \mathfrak{g}$, put

$$\phi(X) = \max_{1 \leq i \leq r} |p_i(X)|^{1/d_i}.$$

Then ϕ is a G-invariant continuous function from $\mathfrak{g}$ to $\mathbb{R}$ and $\phi(X) \geq 0$. Moreover,

(a) $\phi(X) = 0$ if and only if $X \in \mathcal{N}$ and
(b) $\phi(cX) = |c| \, \phi(X)$ for $c \in \Omega$.

Part (b), which follows from the homogeneity of the λ_i and p_j, is not needed here, but it will be used to prove Lemma 11.6. Finally, ϕ is locally constant on $\mathfrak{g}$ outside of $\mathcal{N}$.

Assuming that the $|p_i|$ remain bounded on U is equivalent to assuming that $\phi(X)$ remains bounded on U. Without loss of generality, $U = U^G$ and U is closed.

Let $\mathfrak{h}_1, \mathfrak{h}_2, \ldots, \mathfrak{h}_p$ be a complete set of Cartan subalgebras of $\mathfrak{g}$ under G. $U_i = U \cap \mathfrak{h}_i$ is compact in $\mathfrak{h}_i$. Hence, for a suitable $t > 0$,

$$\bigcup_{1 \le i \le p} U_i \subset \mathfrak{g}(t).$$

Then $U \subset \left(\mathfrak{g}(t)\right)^G$ (see the proof of Lemma 2.1). □

Now we come to the proof of Theorem 11.3. Let W be the complement of V_0 in S. Then W is an open and compact set, and $W + \Lambda = W$. Moreover, if ϕ is as in the proof of Lemma 11.5, ϕ is locally constant on W. Hence we can choose a lattice $\Lambda_0 \subset \Lambda$ such that $\phi(X + \lambda) = \phi(X)$ for all $X \in W$ and $\lambda \in \Lambda_0$. Fix $t_0 \ge t$ such that, for $c \in \Omega^\times$ and $|c| \ge t_0$, we have $c^{-1} L \subset \Lambda_0$. Then the following lemma is obvious from (b) above.

LEMMA 11.6 (Lemma 26). *Suppose X is an element in $\mathfrak{g}$ such that $|X| \ge t_0$ and $X \notin \Omega V_0$. Then*

$$\phi(X + \lambda) = \phi(X)$$

for $\lambda \in L$.

Let q be the number of elements in the residue field of Ω. For $\nu \ge 0$ ($\nu \in \mathbb{Z}$), put

$$\Phi_\nu(X) = \begin{cases} 1 & \text{if } \phi(X) \le q^\nu, \\ 0 & \text{otherwise} \end{cases}$$

for $X \in \mathfrak{g}$. Then Φ_ν is a locally constant function on $\mathfrak{g}$.

LEMMA 11.7 (Lemma 27). *Fix $t_1 \ge t_0$ and $T \in J(V_0, t_1, L)$. Put $T_\nu = \Phi_\nu \cdot T$ for $\nu \ge 0$. Then $T_\nu \in J_0(V_0, t_1, L)$.*

PROOF. $\text{Supp}(\Phi_\nu)$ is a G-invariant set on which ϕ remains bounded. Therefore, by condition (2) of Lemma 11.5, there exists a compact set ω in $\mathfrak{g}$ such that $\text{Supp}(\Phi_\nu) \subset \omega^G$. Therefore $\text{Supp}(\Phi_\nu \cdot T) \subset \omega^G$. This implies $\Phi_\nu \cdot T \in J_0$. Now suppose $X \in \mathfrak{g}$, $|X| \ge t_1$, and $X \notin \Omega V_0$. Then $T_\nu(f_X) = \Phi_\nu(X) \cdot T(f_X)$ by Lemma 11.6 and therefore

$$T_\nu(f_X) = 0$$

since $T \in J(V_0, t_1, L)$. This proves that $T_\nu \in J_0(V_0, t_1, L)$. □

COROLLARY 11.8. *If $C(V, t, L)$ holds, then $j_L J(V_0, t_1, L) = j_L J_0(V_0, t_1, L)$ for $t_1 \ge t_0$.*

PROOF. Since $C(V, t, L)$ implies $C(V_0, t_1, L)$ for $t_1 \ge t$ and $V_0 \subset V$, we conclude from Theorem 11.1 that $\dim j_L J(V_0, t_1, L) < \infty$. Now fix $T \in J(V_0, t_1, L)$, and put $T_\nu = \Phi_\nu \cdot T$. It is obvious that for any fixed $f \in \mathcal{D}$, we must have $T_\nu(f) = T(f)$ for all ν sufficiently large. Therefore

$$T(f) = \lim_{\nu \to \infty} T_\nu(f)$$

for $f \in C_c(\mathfrak{g}/L)$. This shows that $j_L T_\nu$ converges to $j_L T$ in the finite dimensional space $j_L J(V_0, t_1, L)$ as $\nu \to \infty$. Clearly this implies the corollary and hence also proves Theorem 11.3. $\qquad\square$

11.3. Reduction of the condition $C(V, t, L)$. Let L be a lattice in $\mathfrak{g}$. For $X \in \mathfrak{g}$ define f_X as before (§11.1) and let $C(L)$ denote the following condition on L.

CONDITION $C(L)$.   *Given $X_0 \in \mathcal{N} \cap S$, we can choose a neighborhood W of X_0 in S and a number $t \geq 1$ with the following properties. Fix $X \in \Omega W$ with $|X| \geq t$ and define*
$$\phi_\alpha = \int_G \alpha(x) \cdot (f_X)^x \, dx$$
where $\alpha \in C_c^\infty(G)$. Then we can choose α such that

(1) $\int_G \alpha(x) \, dx \neq 0$,
(2) $\phi_\alpha \in C_c(\mathfrak{g}/L)$, *and*
(3) $\mathrm{Supp}(\phi_\alpha) \subset \mathfrak{g}(t')$ *for some t' such that $0 < t' < |X|$.*

LEMMA 11.9 (Lemma 28). *Suppose $C(L)$ holds. Then we can choose a neighborhood V of $\mathcal{N} \cap S$ in S and a number $t > 0$ such that $C(V, t, L)$ holds.*

PROOF. This is a simple consequence of the fact that $\mathcal{N} \cap S$ is compact. $\quad\square$

12. First step in the proof of Howe's Theorem

Assuming a result found in §13, this section provides a proof of Howe's Theorem.

THEOREM 12.1 (Howe's Theorem). *Let ω be a compact subset and L a lattice in $\mathfrak{g}$. Then*
$$\dim j_L J(\omega) < \infty.$$

It follows from Remark 10.7 that, without loss of generality, we may assume that L is well adapted.

LEMMA 12.2. *For all compact subsets ω in $\mathfrak{g}$ there exists a lattice L_1 such that $\omega^G \subset L_1 + \mathcal{N}$.*

PROOF. Fix a p-pair (P, A) with $P = MN$ and let A^+ be the set of all $a \in A$ such that $|\xi_\alpha(a)| \geq 1$ for all positive roots α of (P, A) and their associated characters ξ_α. Then by Bruhat-Tits
$$G = KFA^+K$$
where K is a good maximal compact open subgroup and F is a finite subset of M. Let $\bar{P} = M\bar{N}$ be the p-subgroup opposite to P. Then
$$\mathfrak{g} = \bar{\mathfrak{n}} + \mathfrak{m} + \mathfrak{n}$$
where the sum is direct. We can choose compact subsets ω_1, ω_2 and ω_3 in $\bar{\mathfrak{n}}, \mathfrak{m}$, and $\mathfrak{n}$, respectively, such that
$$\omega^{FK} \subset \omega_1 + \omega_2 + \omega_3.$$

Then
$$\omega^G \subset (\omega_1 + \omega_2 + \omega_3)^{KA^+}.$$
Now $\omega_2^{A^+} = \omega_2$, $\omega_3^{A^+} \subset \mathfrak{n}$, and $\omega_1^{A^+}$ is contained in a compact subset of $\bar{\mathfrak{n}}$. Hence we can choose a compact subset ω_4 in $\mathfrak{g}$ such that
$$(\omega_1 + \omega_2 + \omega_3)^{A^+} \subset \omega_4 + \mathfrak{n}.$$
Therefore
$$\omega^G \subset \omega_4^K + \mathcal{N}. \quad \square$$

Note that $L_1 + \mathcal{N}$ is closed and so $\mathrm{cl}(\omega^G) \subset L_1 + \mathcal{N}$. We may replace L_1 by $L_1 + L$ so that $L \subset L_1$.

LEMMA 12.3. *There exists $t > 0$ and a neighborhood V of $\mathcal{N} \cap S$ in S such that $J(\omega) \subset J(V, t, L)$.*

PROOF. Let $V = (\mathcal{N} \cap S) + \Lambda$ where $|\Lambda| < 1$. Fix a number $t > |L_1|$ so that
$$cL_1 \subset \Lambda$$
for $c \in \Omega$ with $|c| \leq t^{-1}$.

Suppose $X \in \omega^G \subset L_1 + \mathcal{N}$ and $|X| \geq t$. Choose $\ell \in L_1$ such that $Y = X - \ell \in \mathcal{N}$. Then $|\ell| < t$ and therefore
$$|Y| = |X| \geq t.$$

Choose $c \in \Omega^\times$ such that $|c| = |X|^{-1}$. Then $cY \in \mathcal{N} \cap S$ and therefore $cY \in V$. Note that
$$X = Y + \ell = c^{-1}(cY + c\ell) \subset c^{-1}(cY + \Lambda) \subset c^{-1}(V) \subset \Omega V.$$
So, if $T \in J(\omega)$ and $|X| \geq t'$, then $T(f_X) = 0$ unless $X \in \Omega V$. Hence $T \in J(V, t, L)$. $\quad \square$

LEMMA 12.4. *If $C(L)$ holds, then Theorem 12.1 holds.*

PROOF. Choose a neighborhood V of $\mathcal{N} \cap S$ in S and a number $t > 0$ as guaranteed by Lemma 11.9. Without loss of generality, V and t work for Lemma 12.3 also. Thus, by Theorem 11.1, $\dim j_L J(\omega) \leq \dim j_L J(V, t, L) < \infty$. $\quad \square$

13. Completion of the proof of Howe's Theorem

13.1. Reduction to Lemma 13.5.

THEOREM 13.1. *Let L be a well adapted lattice in $\mathfrak{g}$. Fix $X_0 \in \mathcal{N} \cap S$. $C(L)$ holds for X_0.*

We assume L is adapted to (K, A_0) where A_0 is a maximal split torus in G and K is a good maximal compact open subgroup of G corresponding to A_0. Let $X_1, X_2, \ldots, X_n$ be a base as in Definition 10.6.

Complete X_0 to (H_0, X_0, Y_0) as usual by Jacobson-Morosow. Every eigenvalue of $\mathrm{ad}\, H_0$ lies in $\mathbb{Z}$. For $r \in \mathbb{Z}$ let $\mathfrak{g}_r$ be the subspace of $\mathfrak{g}$ consisting of all $Z \in \mathfrak{g}$ such that $[H_0, Z] = rZ$. Put $\mathfrak{n} = \sum_{r \geq 1} \mathfrak{g}_r$, $\mathfrak{m} = \mathfrak{g}_0$, and $\bar{\mathfrak{n}} = \sum_{r \geq 1} \mathfrak{g}_{-r}$. Let $P = N_G(\mathfrak{n})$. Then P is a parabolic subgroup of G. Moreover, $P = MN$ where $M = C_G(H_0)$ and N is the radical of P. Furthermore, $\mathfrak{m}$ and $\mathfrak{n}$ are the Lie algebras of M and N, respectively.

Let A be the maximal split torus lying in the center of M. Then by standard arguments we can choose $k \in K$ and $n \in N$ such that $A^{kn} \subset A_0$. Then A^{kn} is diagonal with respect to $(X_1, \ldots, X_n)$. Put $X_i' = \mathrm{Ad}(k^{-1})X_i$. Since $L^{k^{-1}} = L$,

$$L = \sum_{1 \leq i \leq n} RX_i'$$

and A^n is diagonal with respect to $(X_1', X_2', \ldots, X_n')$. Put $(M', A') = (M^n, A^n)$, $\mathfrak{m}' = \mathfrak{m}^n$, and $\bar{\mathfrak{n}}' = \bar{\mathfrak{n}}^n$. Then $P = M'N$. Put $H_0' = H_0^n$. Then $\mathrm{ad}(H_0')$ is diagonal with respect to $X_1', \ldots, X_n'$. Hence

$$L = L \cap \bar{\mathfrak{n}}' + L \cap \mathfrak{m}' + L \cap \mathfrak{n}.$$

Put $(H_0', X_0', Y_0') = (H_0, X_0, Y_0)^n$ and, as usual, let Γ be the one-parameter subgroup of G corresponding to $\Omega H_0'$. Let ξ denote the rational character of Γ such that $\gamma.X_0' = \xi(\gamma)X_0'$ for $\gamma \in \Gamma$. Since $X_0 \in \mathfrak{n}$, we may choose $\gamma \in \Gamma$ such that

$$(13.1) \qquad\qquad |\xi(\gamma)| < 1 \text{ and } |X_0^\gamma| < |X_0| = 1.$$

Corresponding to the direct sum $\mathfrak{g} = \mathfrak{n} + (\mathfrak{m}' + \bar{\mathfrak{n}}')$, let p_1 and p_2 denote the projection of $\mathfrak{g}$ on $\mathfrak{n}$ and $\mathfrak{m}' + \bar{\mathfrak{n}}'$, respectively. Put $L_i = p_i L$, then $L_1 = L \cap \mathfrak{n}$ and $L_2 = L \cap \mathfrak{m}' + L \cap \bar{\mathfrak{n}}'$. Hence

$$(13.2) \qquad\qquad L_2 \subset L_2^\gamma \subset L^\gamma.$$

LEMMA 13.2. We have $\mathfrak{n} \subset [X_0, \mathfrak{g}]$.

PROOF. This is obvious from the representation theory of (H_0, X_0, Y_0). $\square$

COROLLARY 13.3. We have $\mathfrak{n} \subset [X_0^\gamma, \mathfrak{g}]$.

PROOF. We have $\mathfrak{n} = \mathfrak{n}^\gamma \subset [X_0^\gamma, \mathfrak{g}]$. $\square$

Fix a subspace U of $\mathfrak{g}$ such that

$$u \mapsto [u, X_0^\gamma]$$

from U onto $\mathfrak{n}$ is a bijection. For any $Z \in \mathfrak{g}$, consider the mapping

$$\beta_Z : u \mapsto p_1 [u, Z]$$

of U into $\mathfrak{n}$. Then β_Z depends linearly on Z, and $\beta_{X_0^\gamma}$ is a bijection. Hence we can find a lattice Λ_0 in $\mathfrak{g}$ such that

(1) $|\Lambda_0| < 1$ and $|\Lambda_0^\gamma| < |X_0^\gamma|$ and
(2) there exists a number $\beta_0 > 0$ such that $|\beta_Z(u)| \geq \beta_0 |u|$ for $Z \in (X_0 + \Lambda_0)^\gamma$ and $u \in U$.

Put

$$K_\gamma = K \cap K^\gamma.$$

Then

(13.3) $$L^{k\gamma} = L^\gamma$$

for all $k \in K_\gamma$.

LEMMA 13.4. *For all $Z \in \mathfrak{g}$, let $K_\gamma(Z)$ denote the set of all $k \in K_\gamma$ such that $Z^k - Z \in L$. Then $K_\gamma(Z)$ is an open subgroup of K that depends only on the coset $Z + L$.*

PROOF. The group K_γ operates on $\mathfrak{g}/L$, and $K_\gamma(Z)$ is the stabilizer of $Z + L$ in K_γ under this action. Hence, $K_\gamma(Z)$ is a subgroup of K_γ which depends only on the coset $Z + L$. Clearly, $K_\gamma(Z)$ is open. $\square$

LEMMA 13.5. *Put $W = X_0 + \Lambda$ where Λ is a lattice in $\mathfrak{g}$ such that $\Lambda \subset \Lambda_0$. Then if Λ is sufficiently small, we can choose $t_0 \geq 1$ so that the following conditions hold. If $X \in \Omega W$ and $|X| \geq t_0$, the mapping*

$$\beta \colon k \mapsto p_1(X^{k\gamma} - X^\gamma)$$

of $K_\gamma(X^\gamma)$ into L_1 is surjective. Moreover, it is everywhere submersive.

Assuming this lemma, we shall prove our main result (Theorem 13.1) and thus complete the proof of Theorem 12.1. Fix Λ, W, and t_0 as in Lemma 13.5.

Fix $t \geq t_0$ so large that

(13.4) $$c^{-1}L \subset \Lambda \cap \Lambda^\gamma$$

if $c \in \Omega$ and $|c| \geq t$. Now fix $X \in \Omega W$ with $|X| \geq t$. Choose $c \in \Omega$ such that $X \in cW$. Then $|c| = |X| \geq t$. Hence

$$c^{-1}(X + L) \subset W + c^{-1}L \subset W + \Lambda = W.$$

Put $K_0 = K_\gamma(X^\gamma)$, and consider

$$\phi_\alpha = \int_{K_0} \alpha(k) \cdot f_X^{k\gamma} \, dk$$

where $\alpha \in C_c^\infty(K_0)$. It follows from equation (13.3) that $f_X^{k\gamma}$ is the characteristic function of $X^{k\gamma} + L^{k\gamma} = X^{k\gamma} + L^\gamma$. (Recall that $K_0 \subset K_\gamma$.) Let F be the characteristic function of L, then F^γ is the characteristic function of L^γ. Hence $f_X^{k\gamma}(Z) = F^\gamma(Z - X^{k\gamma})$ for $Z \in \mathfrak{g}$. Therefore

$$\phi_\alpha(Z) = \int_{K_0} \alpha(k) \cdot F^\gamma(Z - X^{k\gamma}) \, dk$$

for $Z \in \mathfrak{g}$.

Now

(13.5) $$X^{k\gamma} = X^\gamma + (X^{k\gamma} - X^\gamma) = X^\gamma + p_1(X^{k\gamma} - X^\gamma) + p_2(X^{k\gamma} - X^\gamma).$$

But $X^{k\gamma} - X^\gamma \in L = p_1 L + p_2 L$ and by equation (13.2) $p_2 L \subset L^\gamma$. Hence $F^\gamma(Z - X^{k\gamma}) = F^\gamma\big(Z - X^\gamma - \beta(k)\big)$ for $k \in K_0$ where β is the mapping

$$\beta \colon k \mapsto p_1(X^{k\gamma} - X^\gamma).$$

By Lemma 13.5, β is a surjective submersion of K_0 onto L_1. Hence, by [**21**, Theorem 11, p. 49], there exists a surjective linear mapping $\alpha \mapsto f_\alpha$ from $C_c^\infty(K_0)$ onto $C_c^\infty(L_1)$ such that

$$\int_{L_1} f_\alpha(v) \cdot \theta(v)\, dv = \int_{K_0} \alpha(k) \cdot \theta\big(\beta(k)\big)\, dk$$

for all $\theta \in C(L_1)$. Here dv is the Haar measure on L_1. We may choose α so that $f_\alpha = 1$. Putting $\theta = 1$ we have

$$\int_{K_0} \alpha(k)\, dk = \int_{L_1} dv \neq 0,$$

and, putting $\theta(v) = F^\gamma(Z - X^\gamma - v)$, we have

$$\phi_\alpha(Z) = \int_{K_0} \alpha(k) \cdot F^\gamma\big(Z - X^\gamma - \beta(k)\big)\, dk$$

$$= \int_{L_1} F^\gamma(Z - X^\gamma - v)\, dv.$$

If $\lambda \in L$, then $\lambda = \lambda_1 + \lambda_2$ where $\lambda_i = p_i \lambda$. So

$$\phi_\alpha(Z - \lambda) = \int_{L_1} F^\gamma(Z - \lambda_1 - \lambda_2 - X^\gamma - v)\, dv.$$

Now $\lambda_2 \in p_2 L \subset L^\gamma$ and $\lambda_1 \in L_1$. Hence $\phi_\alpha(Z - \lambda) = \phi_\alpha(Z)$. This proves that $\phi_\alpha \in C_c(\mathfrak{g}/L)$.

It is obvious that if $Z \in \mathrm{Supp}(\phi_\alpha)$, then $Z \in X^{k\gamma} + L^\gamma$ (as in equation (13.5)). Since $X \in cW = c(X_0 + \Lambda)$ and $|c| \geq t$, it follows from equation (13.4) that

$$c^{-1}L \subset \Lambda \cap \Lambda^\gamma.$$

Hence

$$c^{-1}(L + L^\gamma) \subset \Lambda^\gamma$$

and so

(13.6) $Z \in X^\gamma + L + L^\gamma \subset c(X_0 + \Lambda)^\gamma + c\Lambda^\gamma \subset c(X_0 + \Lambda)^\gamma.$

It follows from the requirements on page 64 for Λ_0 and equation (13.1) that

$$|Z| = |c||X_0^\gamma| < |c| = |X|.$$

This completes the proof of Theorem 13.1 (modulo the proof of Lemma 13.5).

13.2. Proof of Lemma 13.5. It remains to prove Lemma 13.5. The proof requires a great deal of 'book keeping' and so we place our assumptions at the beginning.

Choose $\varepsilon \in (0,1)$. Then $\mathfrak{g}(\varepsilon)$ is a lattice in $\mathfrak{g}$. By choosing ε sufficiently small we can assume that

(1) $[X_1, X_2] \in \mathfrak{g}(\varepsilon)$ for $X_i \in \mathfrak{g}(\varepsilon)$,
(2) $\mathfrak{g}(\varepsilon) \subset L$,
(3) $\mathfrak{g}(\varepsilon)$ is admissible, and $K_\varepsilon = \exp(\mathfrak{g}(\varepsilon))$ is an open subgroup of K_γ, and
(4) there exists a real number $a_3 > 0$ such that $\big|\mathrm{Ad}(\exp Z)Y - Y - [Z,Y]\big| \leq a_3 |Z|^2 |Y|$ for $Z \in \mathfrak{g}(\varepsilon)$ and $Y \in \mathfrak{g}$.

Choose $a_1, a_2 \in \mathbb{R}$ with $a_i > 0$ such that for all $Z \in \mathfrak{g}$

$$(13.7) \qquad\qquad |p_i Z| \leq a_i |Z|.$$

Choose a lattice $\Lambda \subset \Lambda_0$ such that

$$(13.8) \qquad\qquad \frac{a_2}{\beta_0} |L_1||\Lambda^\gamma| \leq \varepsilon.$$

Choose $t_0 \geq 1$ so that

(1) $c^{-1}L \subset \Lambda^\gamma$ whenever $\big|c^{-1}\big| \leq t_0^{-1}$,
(2) $|L_1|\beta_0^{-1} t_0^{-1} \leq \varepsilon$,
(3) $a_3 |L_1|^2 \beta_0^{-2} t_0^{-1} \leq \varepsilon$, and
(4) $a_3 a_1 |L_1| \beta_0^{-2} t_0^{-1} \leq \frac{1}{2}$.

Fix $X \in \Omega W = \Omega(X_0 + \Lambda)$ such that $|X| \geq t_0$. Then $X \in c\,(X_0 + \Lambda)$ with $|c| = |X| \geq t_0$. If $k \in K_\gamma(X^\gamma) = K_0$, then, as in equation (13.6),

$$X^{k\gamma} \in c\,(X_0 + \Lambda)^\gamma.$$

This implies that

$$|\beta_{X^{k\gamma}}(u)| \geq |c| \cdot \beta_0 \cdot |u|$$

for all $u \in U$ (equation (13.2)). Hence, the mapping

$$u \mapsto \beta_{X^{k\gamma}}(u)$$

of U to $\mathfrak{n}$ is bijective. Since K_0 is open, this proves that the mapping

$$\beta \colon k \mapsto p_1(X^{k\gamma} - X^\gamma)$$

from K_0 to L_1 is everywhere submersive. We now want to show that β is surjective onto L_1.

Choose $v \in L_1$.

Let $Z \in c\,(X_0 + \Lambda)^\gamma$ with c as above. Note that $|Z| = |c X_0^\gamma| < |c|$ since $|\Lambda^\gamma| \leq |\Lambda_0^\gamma| < 1$ and $|X_0^\gamma| < 1$ by equation (13.1). Furthermore $\big|\gamma^{-1}.Z\big| = |c|$ since $|\Lambda| \leq |\Lambda_0| < 1$. So, by the conditions for Λ_0 on page 64,

$$(13.9) \qquad\qquad \big|\beta_Z^{-1} v\big| \leq \frac{|v|}{\beta_0 \, |c|} = \frac{|v|}{\beta_0 \, \big|\gamma^{-1}.Z\big|}$$

since $v \in \mathfrak{n} = p_1\mathfrak{g}$. Moreover

$$\left[\beta_Z^{-1}v, Z\right] = p_1\left[\beta_Z^{-1}v, Z\right] + p_2\left[\beta_Z^{-1}v, Z\right] = v + p_2\left[\beta_Z^{-1}v, Z\right].$$

Now $Z = c\left(X_0^\gamma + \lambda^\gamma\right)$ where $\lambda \in \Lambda$. Since $[U, X_0^\gamma] = \mathfrak{n}$, we have $p_2\left[\beta_Z^{-1}v, X_0^\gamma\right] = 0$. Hence

$$\left|p_2\left[\beta_Z^{-1}v, Z\right]\right| = \left|c\,p_2\left[\beta_Z^{-1}v, \lambda^\gamma\right]\right| \le \left|\gamma^{-1}.Z\right| \cdot a_2 \cdot \left|\beta_Z^{-1}v\right| \cdot \left|\lambda^\gamma\right|$$
$$\le \frac{a_2\,|v|\,|\Lambda^\gamma|}{\beta_0}.$$

Therefore, since $v \in L_1$,

(13.10)
$$\left|\left[\beta_Z^{-1}v, Z\right] - v\right| \le \frac{a_2\,|v|\,|\Lambda^\gamma|}{\beta_0} \le \varepsilon$$

by equation (13.8), and

(13.11)
$$\left|\beta_Z^{-1}v\right| \le \frac{|v|}{\beta_0\left|\gamma^{-1}.Z\right|} \le \varepsilon.$$

Put $k(Z, v) = \exp(\beta_Z^{-1}v)$ and $K_\varepsilon(Z) = K_\varepsilon \cap K_\gamma(Z)$.

LEMMA 13.6. *We have $k(Z, v) \in K_\varepsilon(Z)$*

PROOF. Put $k = k(Z, v)$ and $u = \beta_Z^{-1}v$. Since $u \in \mathfrak{g}(\varepsilon)$ (equation (13.11)), it is enough to show that $k \in K_\gamma(Z)$. From equation (13.10) we have that $|[u, Z] - v| \le \varepsilon$ which implies $[u, Z] \in L$. So it is enough to show that $\left|Z^k - Z - [u, Z]\right| \le \varepsilon$. But, by condition (13.2) on page 67 for t_0,

$$\left|Z^k - Z - [u, Z]\right| \le a_3\,|u|^2\,|Z| \le \frac{a_3\,|v|^2\,|Z|}{\beta_0^2\left|\gamma^{-1}.Z\right|^2}$$
$$\le \frac{a_3\,|v|^2}{\beta_0^2\,|c|} \le \frac{a_3\,|v|^2}{\beta_0^2\,t_0} \le \varepsilon. \quad \square$$

Continue the notation of the lemma. Note that since $p_1\,[u, Z] = v$, we have $p_1(Z^k - Z) - v = p_1(Z^k - Z - [u, Z])$. Hence, since $|Z| < \left|\gamma^{-1}.Z\right|$,

(13.12)
$$\left|p_1(Z^k - Z) - v\right| \le \frac{a_3\,a_1\,|v|^2}{\beta_0^2\left|\gamma^{-1}.Z\right|}.$$

We shall show that there exists an element $k \in K_\varepsilon(X^\gamma)$ such that $v = p_1(X^{k\gamma} - X^\gamma)$. We define elements $v_n \in L_1$, $Z_n \in X^\gamma + L$, and $k_n \in K_\varepsilon(X^\gamma)$ inductively as follows. Put

$$v_0 = v, \ Z_0 = X^\gamma, \text{ and } k_0 = 1.$$

Put

$$k_n = k(Z_{n-1}, v_{n-1}), \ Z_n = Z_{n-1}^{k_n}, \text{ and } v_n = v_{n-1} - p_1(Z_n - Z_{n-1})$$

for $n \ge 1$. By the induction hypothesis, $Z_{n-1} \in X^\gamma + L$ and $v_{n-1} \in L_1$. So, from Lemma 13.6, we have $k_n \in K_\varepsilon(Z_{n-1}) = K_\varepsilon(X^\gamma)$, and therefore

$$Z_n \in Z_{n-1} + L = X^\gamma + L.$$

It follows that
$$p_1(Z_n - Z_{n-1}) \in L_1,$$
and therefore $v_n \in L_1$.

Now put $k'_n = k_n k_{n-1} k_{n-2} \cdots k_2 k_1$ for $n \geq 1$. Then
$$Z_n = X^{k'_n \gamma},$$
and therefore
$$X^{k'_n \gamma} - X^\gamma = Z_n - Z_0 = \sum_{1 \leq r \leq n} (Z_r - Z_{r-1}).$$
Hence
$$p_1(X^{k'_n \gamma} - X^\gamma) = \sum_{1 \leq r \leq n} p_1(Z_r - Z_{r-1}) = \sum_{1 \leq r \leq n} (v_{r-1} - v_r)$$
$$= v - v_n.$$

Note that
$$Z_{n-1} \in X^\gamma + L \subset c\,(X_0 + \Lambda)^\gamma,$$
and so $\left|\gamma^{-1}.Z_{n-1}\right| = |c| = |X| \geq t_0$. From this fact and equation (13.12) it follows that
$$|v_n| = |v_{n-1} - p_1(Z_n - Z_{n-1})| \leq \frac{a_1\,a_3\,|v_{n-1}|^2}{\beta_0^2\,\left|\gamma^{-1}.Z_{n-1}\right|}$$
$$\leq \frac{a_1\,a_3\,|v_{n-1}|^2}{\beta_0^2\,t_0} \leq \frac{a_1\,a_3\,|L_1|\,|v_{n-1}|}{\beta_0^2\,t_0}$$
$$\leq \frac{|v_{n-1}|}{2}.$$

Therefore $|v_n| \to 0$ as $n \to \infty$. This shows that
$$\lim_{n \to \infty} p_1(X^{k'_n \gamma} - X^\gamma) = v.$$

However, $K_\varepsilon(X^\gamma)$ is a compact group. Hence by choosing a subsequence we may assume that $k_n \to k'$ where $k' \in K_\varepsilon(X^\gamma)$. Then
$$p_1(X^{k'\gamma} - X^\gamma) = v.$$

This completes the proof of Lemma 13.5.

Part III. Theory on the group

14. Representations of compact groups

Let K be a compact group. We denote by $\mathcal{E}(K)$ the set of all equivalence classes of irreducible representations of K. For $\underline{d} \in \mathcal{E}(K)$, let $\xi_{\underline{d}}$ denote the character and $d(\underline{d})$ the degree of $\underline{d}$. For a finite subset F of $\mathcal{E}(K)$, put

$$\xi_F = \sum_{\underline{d} \in F} \xi_{\underline{d}},$$

and denote by $\mathcal{A}(F)$ the subspace of $C(K)$ spanned by the left and right translates of ξ_F. (Note that $\mathcal{A}(F) = \sum_{\underline{d} \in F} \mathcal{A}(\underline{d})$, and $\mathcal{A}(\underline{d})$ may be characterized as the vector space generated by the matrix coefficients of $\underline{d}$.)

Suppose L is a closed subgroup of K. For $\underline{d} \in \mathcal{E}(K)$ and $\delta \in \mathcal{E}(L)$, we write $[\underline{d}: \delta]$ for the multiplicity of δ in the restriction of $\underline{d}$ to L.

Let G be a topological group. Suppose K_1 and K_2 are two compact subgroups of G. For finite subsets F_i of $\mathcal{E}(K_i)$, we say that F_1 and F_2 interact if there exist elements $\underline{d}_i \in F_i$ and $\delta \in \mathcal{E}(K_1 \cap K_2)$ such that $[\underline{d}_1: \delta] \geq 1$ and $[\underline{d}_2: \delta] \geq 1$. Put

$$[F_1: F_2] = \int_{K_1 \cap K_2} \bar{\xi}_{F_1}(k) \cdot \xi_{F_2}(k)\, dk$$

where dk is the normalized Haar measure of $K_1 \cap K_2$. For $c \in \mathbb{C}$, we denote by $\bar{c}$ the complex conjugate of c. For a function $f: G \to \mathbb{C}$, we define $\bar{f}$ as follows. If $x \in G$, then $\bar{f}(x) = \bar{c}$ where $c = f(x)$. The following lemma is an immediate consequence of the orthogonality relations for characters of compact groups.

LEMMA 14.1 (Lemma 30). *We have $[F_1: F_2]$ is a nonnegative integer. Moreover, we have $[F_1: F_2] = [F_2: F_1]$ and $[F_1: F_2] \neq 0$ if and only if F_1 and F_2 interact.*

If $x \in G$ and S is any subset of G, we put $^xS = xSx^{-1}$. Let K be a compact subgroup of G. Then $k \mapsto {}^xk$ for $k \in K$ is an isomorphism of K onto ^xK. Hence, it defines a bijection

$$\underline{d} \mapsto {}^x\underline{d}$$

of $\mathcal{E}(K)$ onto $\mathcal{E}(^xK)$.

Suppose F_1 and F_2 are as in Lemma 14.1 and $x \in G$. We say that x intertwines F_2 with F_1 if $[F_1: {}^xF_2] \neq 0$. Let S be a subset of G. We say that S intertwines F_2 with F_1 if $[F_1: {}^xF_2] \neq 0$ for some $x \in S$.

LEMMA 14.2 (Lemma 31). *If* $[F_1 : F_2] = 0$, *then* $\int_{K_1 \cap K_2} \bar{f}_1(k) \cdot f_2(k)\, dk = 0$ *for all* $f_i \in \mathcal{A}(F_i)$.

PROOF. This is an easy consequence of orthogonality since F_1 and F_2 do not interact. $\qquad\square$

COROLLARY 14.3. *Fix* $x \in G$ *and let* f *be a complex-valued function on* $K_1 x K_2$ *such that the function*

$$(k_1, k_2) \mapsto f(k_1 x k_2)$$

on $K_1 \times K_2$ *lies in* $\mathcal{A}(F_1) \otimes \mathcal{A}(F_2)$. *Then if* $f(x) \neq 0$, x *intertwines* F_2 *with* F_1.

PROOF. Replacing K_2 by ${}^x K_2$ we may assume $x = 1$. Let f_1, f_2, and g be the restrictions of f on K_1, K_2, and $K = K_1 \cap K_2$, respectively. Then $f_i \in \mathcal{A}(F_i)$ and so f_i is a continuous function. Moreover, $f_1 = f_2 = g$ on K. Since $g(1) = f(1) \neq 0$, we have

$$0 \neq \int_K |g(k)|^2\, dk = \int_K \bar{f}_1(k) \cdot f_2(k)\, dk.$$

This shows that $[F_1 : F_2] \neq 0$. $\qquad\square$

Let K_0, K_1, and K_2 be three compact subgroups of G such that $K_3 = K_1 \cap K_2$ is open in both K_1 and K_2.

LEMMA 14.4 (Lemma 32). *Fix* $\underline{d}_0 \in \mathcal{E}(K_0)$ *and* $\underline{d}_1 \in \mathcal{E}(K_1)$. *Let* F_2 *be the set of all* $\underline{d}_2 \in \mathcal{E}(K_2)$ *such that* $[\underline{d}_1 : \underline{d}_2] \neq 0$. *Then the condition* $[\underline{d}_0 : \underline{d}_1] \neq 0$ *implies* $[\underline{d}_0 : F_2] \neq 0$.

The proof of this lemma requires some preparation.

Note that F_2 is a finite set. Extend any function on K_3 to a function on K_2 by defining it to be 0 outside K_3. Let F_3 be the set of all elements $\underline{d}_3 \in \mathcal{E}(K_3)$ such that $[\underline{d}_1 : \underline{d}_3] \neq 0$.

LEMMA 14.5. *We have* $\mathcal{A}(F_3) \subset \mathcal{A}(F_2)$.

Before proving this lemma, we prove:

LEMMA 14.6. *Lemma 14.5 implies Lemma 14.4.*

PROOF. Since $[\underline{d}_0 : \underline{d}_1] \neq 0$, we have

$$0 \neq \int_{K_0 \cap K_1} \bar{\xi}_{\underline{d}_0}(k) \cdot \xi_{\underline{d}_1}(k)\, dk.$$

Hence, we can choose $k_0 \in K_0 \cap K_1$ such that

$$\int_{K_0 \cap K_3} \bar{\xi}_{\underline{d}_0}(k_0 k) \cdot \xi_{\underline{d}_1}(k_0 k)\, dk \neq 0.$$

Let f denote the function on K_3 given by

$$f(k) = \xi_{\underline{d}_1}(k_0 k)$$

for $k \in K_3$. Then $f \in \mathcal{A}(F_3) \subset \mathcal{A}(F_2)$. The function $g \colon k \mapsto \xi_{\underline{d}_0}(k_0 k)$ for $k \in K_0$ lies in $\mathcal{A}(\underline{d}_0)$. Hence

$$0 \neq \int_{K_0 \cap K_2} \bar{g}(k) \cdot f(k) \, dk.$$

So $[\underline{d}_0 \colon F_2] \neq 0$. □

Now we come to Lemma 14.5. Fix $\underline{d}_3 \in F_3$ and let $f \in \mathcal{A}(\underline{d}_3)$. Then

$$f = \sum_{\underline{d} \in \mathcal{E}(K_2)} d(\underline{d}) f * \xi_{\underline{d}}$$

where the convolution is with respect to K_2. But $f * \xi_{\underline{d}} = 0$ unless $\underline{d}$ occurs in $\mathrm{Ind}_{K_3}^{K_2} \underline{d}_3$ (i.e., unless $\underline{d}$ interacts with $\underline{d}_3$). But if $\underline{d}$ interacts with $\underline{d}_3$, then $\underline{d} \in F_2$. Hence

$$f = \sum_{\underline{d} \in F_2} d(\underline{d}) \cdot f * \xi_{\underline{d}} \in \mathcal{A}(F_2). \quad □$$

LEMMA 14.7. *Notation as above. Fix a finite subset F_1 of $\mathcal{E}(K_1)$, and let F_2 be the set of all $\underline{d} \in \mathcal{E}(K_2)$ such that $[F_1 \colon \underline{d}] \neq 0$. For any $x \in G$, let $\mathcal{E}_{x,F_i}(K_0)$ be the set of all $\underline{d}_0 \in \mathcal{E}(K_0)$ such that x intertwines $\underline{d}_0$ to F_i. Then*

$$\mathcal{E}_{x,F_1}(K_0) \subset \mathcal{E}_{x,F_2}(K_0).$$

PROOF. Note that F_2 is a finite set.

It is enough to consider the case $x = 1$. Fix $\underline{d}_0 \in \mathcal{E}_{1,F_1}(K_0)$. Then $[\underline{d}_0 \colon \underline{d}_1] \neq 0$ for some $\underline{d}_1 \in F_1$. By Lemma 14.4, we have that there exists $\underline{d}_2 \in F_2$ such that $[\underline{d}_0 \colon \underline{d}_2] \neq 0$. Hence $\underline{d}_0 \in \mathcal{E}_{1,F_2}(K_0)$. □

Let K be an open compact subgroup of G and dk the normalized Haar measure on K. Assume that G is totally disconnected.

LEMMA 14.8 (Lemma 33). *Let Θ be a distribution of K. Put*

$$\Theta_{\underline{d}} = d(\underline{d}) \cdot \Theta * \xi_{\underline{d}}$$

where $\underline{d} \in \mathcal{E}(K)$ and $*$ represents convolution on K with respect to dk. Then

$$\Theta = \sum_{\underline{d} \in \mathcal{E}(K)} \Theta_{\underline{d}}.$$

Moreover, if Θ is K invariant,

$$\Theta_{\underline{d}} = \Theta(\bar{\xi}_{\underline{d}}) \cdot \xi_{\underline{d}}.$$

PROOF. This is an immediate consequence of the Plancherel formula for K. □

15. Admissible distributions

Let G be a totally disconnected topological group [20, §§2,3] which we assume to be unimodular. Let Θ be a distribution on an open, G-invariant set U of G, K_0 a compact open subgroup of G, and γ an element of U.

DEFINITION 15.1. We say that Θ is (G, K_0)-admissible at γ if

(1) $\gamma K_0 \subset U$ and
(2) for any open subgroup K of K_0 and $\underline{d} \in \mathcal{E}(K)$,

$$\Theta * \xi_{\underline{d}} = 0$$

on γK_0 unless G intertwines 1_{K_0} with $\underline{d}$.

Here the $*$ denotes convolution, and 1_{K_0} is the class of the trivial representation of K_0.

Suppose Θ is (G, K_0)-admissible at γ and K_1 is an open subgroup of K_0. Then it is easy to deduce from Lemma 14.4 that Θ is (G, K_1)-admissible at every point in γK_0.

We shall say that Θ is G-admissible at γ if it is (G, K_0)-admissible at γ for some open compact subgroup K_0 of G.

By an admissible distribution on G, we will mean a distribution which is G-admissible at every point.

Let π be an irreducible, admissible representation of G on V and Θ_π its character [20, §3]. Let K_0 be an open compact subgroup of G and V_0 the space of all elements in V which are left fixed by K_0. If K_0 is sufficiently small, $V_0 \neq \{0\}$. Then it is easy to verify that Θ_π is (G, K_0)-admissible at every point (see [4]).

16. Statement of the main results

Now let G have the same meaning as in Part I. It operates on itself by conjugation. Our object is to prove the following three theorems.

THEOREM 16.1 (Theorem 19). *Let Θ be a G-invariant distribution on an open G-invariant subset U of G. Let γ be a semisimple element in U. Then if Θ is G-admissible at γ, it coincides with a locally summable function around γ.*

Let M and $\mathfrak{m}$ be the centralizers of γ in G and $\mathfrak{g}$, respectively.

THEOREM 16.2 (Theorems 5 and 20). *Under the above conditions, there exist unique complex numbers c_ξ such that*

$$\Theta(\gamma \exp Y) = \sum_\xi c_\xi \cdot \widehat{\nu}_\xi(Y)$$

for Y sufficiently close to zero in $\mathfrak{m}$. Here ξ runs over all nilpotent M-orbits in $\mathfrak{m}$, ν_ξ is the M-invariant measure[4] on $\mathfrak{m}$ corresponding to ξ, and $\widehat{\nu}_\xi$ is the Fourier transform of ν_ξ on $\mathfrak{m}$.

THEOREM 16.3 (Theorem 1). *Let Θ_π denote the character of an admissible and irreducible representation of G. Then Θ_π is a locally summable function on G which is locally constant on G'. Moreover, the function*

$$|D_G|^{1/2} \cdot \Theta_\pi$$

is locally bounded on G.

Theorems 16.2 and 16.3 will be proved in §21.

17. Recapitulation of Howe's theory

We shall now recall some results of Howe[5] [**24**].

As before let Z denote the maximal split torus contained in the center of G. Also, for $x \in G$, let $D(x) = D_G(x)$ denote the coefficient of t^ℓ in the polynomial

$$\det\big(t - \mathrm{Ad}(x) + 1\big)$$

where t is an indeterminate and $\ell = \mathrm{rank}\, G$.

Choose a $(G, \mathfrak{g})$-admissible G-domain $\mathfrak{g}_0$. Let $G_0 = \exp(\mathfrak{g}_0)$.

Let p be the rational prime dividing the prime $\mathfrak{p}$ of Ω and e the ramification index of $\mathfrak{p}$ over p. Put

$$\beta = \begin{cases} \left\lceil \frac{3e}{2(p-1)} \right\rceil + 1 & \text{if } p \neq 3, \\ e + 1 & \text{if } p = 3. \end{cases}$$

Here $\lceil\ \rceil$ denotes the greatest integer function. Fix an element ϖ in R of order one. Suppose L is a lattice in $\mathfrak{g}$ such that

(1) $L \subset \mathfrak{g}_0$ and
(2) $[L, L] \subset \varpi^\beta L$.

Then $K = \exp L$ is an open compact subgroup of G [**24**].

By an s-lattice[6] we mean a lattice L satisfying the above two conditions. If $p = 2$, we shall further assume that

(1) $L \subset 2\mathfrak{g}_0$ and
(2) $[L, L] \subset 2^3 L$.

Then in every case $K = \exp L$ and $K^{1/2} = \exp(\frac{1}{2}L)$ are compact open subgroups of G, and K is normal in $K^{1/2}$. (Note that $L = \frac{1}{2}L$ and $K = K^{1/2}$ unless $p = 2$.) Hence $K^{1/2}$ operates on K by conjugation and therefore also on $\mathcal{E}(K)$. Let $\mathcal{E}^{1/2}(K)$

---

[4]The centralizer $\mathbf{M}$ of γ in $\mathbf{G}$ is, in general, not connected. But since the connected component of the identity is a subgroup of finite index, this does not present a serious problem. In particular, the $\widehat{\nu}_\xi$ are functions.

[5]For a fuller discussion of this material, see [**4**].

[6]This term is meant to signify that the lattice is small in a technical sense.

denote the set of all $K^{1/2}$-orbits in $\mathcal{E}(K)$. Since $\left[K^{1/2} : K\right] < \infty$, every such orbit is a finite subset of $\mathcal{E}(K)$.

Fix $\underline{d} \in \mathcal{E}^{1/2}(K)$ and let $\underline{d}_1, \underline{d}_2, \dots, \underline{d}_N$ be all the distinct elements of $\mathcal{E}(K)$ lying on the orbit $\underline{d}$. It is clear that $d(\underline{d}_i)$ is independent of i for $1 \leq i \leq N$. Put $d(\underline{d}) = d(\underline{d}_i)$ and

$$\xi_{\underline{d}} = \sum_i \xi_{\underline{d}_i}.$$

We call $d(\underline{d})$ the degree and $\xi_{\underline{d}}$ the character of $\underline{d}$.

Fix an s-lattice in $\mathfrak{g}$, and let L^* denote the dual lattice with respect to $(X, Y) \mapsto \chi\big(B(X, Y)\big)$. Define $K = \exp L$ and $K^{1/2} = \exp(\frac{1}{2}L)$ as above. Then L^* is stable under $\mathrm{Ad}(K^{1/2})$, and therefore $\mathrm{Ad}(K^{1/2})$ operates on $\mathfrak{g}/L^*$.

THEOREM 17.1 (Theorem 21 (Howe)). *There is a bijection*

$$\underline{d} \mapsto \mathcal{O}_{\underline{d}},$$

from $\mathcal{E}^{1/2}(K)$ to the set of all $K^{1/2}$-orbits in $\mathfrak{g}/L^*$, such that for $\lambda \in L$,

$$d(\underline{d}) \cdot \xi_{\underline{d}}(\exp \lambda) = \sum_{X \in \mathcal{O}_{\underline{d}}/L^*} \chi\big(B(X, \lambda)\big)$$

and

$$d(\underline{d}) = [K : K_X]^{1/2}.$$

Here X is any element in $\mathcal{O}_{\underline{d}}$ and K_X is the subgroup of all elements $k \in K$ such that

$$\mathrm{Ad}(k)X \in X + L^*.$$

It is convenient to regard a subset of $\mathfrak{g}/L^*$ as a subset of $\mathfrak{g}$ (by identifying it with its pre-image). This holds, in particular, for $\mathcal{O}_{\underline{d}}$. Suppose L_1 and L_2 are two s-lattices in $\mathfrak{g}$. Put $K_i = \exp(L_i)$.

COROLLARY 17.2 (Howe). *Fix $\underline{d}_i \in \mathcal{E}^{1/2}(K_i)$ and let $\mathcal{O}_i$ denote the $K_i^{1/2}$-orbit in $\mathfrak{g}/L_i^*$ corresponding to $\underline{d}_i$. Then an element $x \in G$ intertwines $\underline{d}_2$ with $\underline{d}_1$ if and only if $\mathcal{O}_1 \cap {}^x\mathcal{O}_2 \neq \emptyset$.*

18. Application to admissible invariant distributions

We return to the notation of §16. In particular, U is as on page 74. Let M be the centralizer of γ in G. Let $\mathfrak{m}$ be the corresponding Lie algebra. Put

$$\mathfrak{q} = \big(\mathrm{Ad}(\gamma) - 1\big)\mathfrak{g}.$$

Since γ is semisimple, $\mathfrak{g} = \mathfrak{m} + \mathfrak{q}$ where the sum is direct. Moreover, $\mathfrak{m}$ and $\mathfrak{q}$ are orthogonal under the bilinear form B, and the restriction of B on $\mathfrak{m}$ is nondegenerate. Hence $\mathfrak{m}$ is reductive.

For $m \in M$, put

$$D_{G/M}(m) = \det\big(\mathrm{Ad}(m) - 1\big)\big|_{\mathfrak{g}/\mathfrak{m}},$$

and let M' be the set of all $m \in M$ where $D_{G/M}(\gamma m) \neq 0$. Put

$$U_M = M' \cap (\gamma^{-1}U).$$

Then U_M is an open and M-invariant neighborhood of the identity in M. The mapping

$$(x, m) \mapsto x\gamma m x^{-1}$$

of $G \times U_M$ into U is submersive [59, Proposition 1], and $U_0 = (\gamma U_M)^G$ is an open and G-invariant neighborhood of γ in U. Hence, there exists a unique, surjective, linear mapping

$$\alpha \mapsto f_\alpha$$

of $C_c^\infty(G \times U_M)$ onto $C_c^\infty(U_0)$ such that

$$\int_{G \times U_M} \alpha(x : m) \cdot F(x\gamma m x^{-1}) \, dx \, dm = \int_U f_\alpha(x) \cdot F(x) \, dx$$

for all $F \in C(G)$. (dx and dm are the Haar measures on G and M, respectively.) Moreover, we conclude from [21, p. 56] that there exists a unique M-invariant distribution θ on U_M such that

$$\Theta(f_\alpha) = \theta(\beta_\alpha)$$

for $\alpha \in C_c^\infty(G \times U_M)$ where, for $m \in M$,

$$\beta_\alpha(m) = \int_G \alpha(x : m) \, dx.$$

Let K_0 be a compact open subgroup of G such that the distribution Θ of Theorem 16.1 is (G, K_0)-admissible at γ. Fix a compact open subgroup M_0 of M such that $M_0 \subset U_M \cap K_0$, and let K be any open subgroup of K_0. Fix $\beta \in C_c^\infty(M_0)$, and define $\alpha \in C_c^\infty(G \times M_0)$ by

$$\alpha(x : m) = \alpha_K(x) \cdot \beta(m)$$

for $x \in G$ and $m \in M$. Here α_K is the characteristic function of K divided by the total measure of K. Then $\beta_\alpha = \beta$ and so

$$\theta(\beta) = \Theta(f_\alpha) = \sum_{\underline{d} \in \mathcal{E}(K)} \int_G f_\alpha(x) \cdot \Theta_{\underline{d}}(x) \, dx$$

from the Plancherel formula for K. Moreover, we observe that

$$\mathrm{Supp}(f_\alpha) \subset (\gamma M_0)^K \subset K_0 \gamma K_0 \subset U.$$

But

$$\int_G f_\alpha(x) \cdot \Theta_{\underline{d}}(x) \, dx = \int_{G \times U_M} \alpha(x : m) \cdot \Theta_{\underline{d}}(x\gamma m x^{-1}) \, dx \, dm$$

$$= \int_{M_0} \beta(m) \cdot \Theta_{\underline{d}}(\gamma m) \, dm$$

since $\Theta_{\underline{d}}$ is K-invariant. Hence, for $\beta \in C_c^\infty(M_0)$,

$$\theta(\beta) = \sum_{\underline{d} \in \mathcal{E}(K)} \int_{M_0} \beta(m) \cdot \Theta_{\underline{d}}(\gamma m)\, dm.$$

We normalize dm so that M_0 has total measure one.

LEMMA 18.1 (Lemma 34). *We have*

$$\theta = \sum_{\delta \in \mathcal{E}(M_0)} \theta(\bar{\xi}_\delta) \cdot \xi_\delta$$

on M_0, and, for $\delta \in \mathcal{E}(M_0)$,

$$\theta(\bar{\xi}_\delta) = \sum_{\underline{d} \in \mathcal{E}(K)} \int_{M_0} \bar{\xi}_\delta(m) \cdot \Theta_{\underline{d}}(\gamma m)\, dm.$$

PROOF. See Lemma 14.8.   □

For any finite subset F of $\mathcal{E}(K)$, put

$$\Theta_F = \sum_{\underline{d} \in F} \Theta_{\underline{d}}.$$

LEMMA 18.2 (Lemma 35). *Let F be a finite subset of $\mathcal{E}(K)$. Fix $\delta \in \mathcal{E}(M_0)$, and suppose $[F : \delta] = 0$. Then*

$$\int_{M_0} \bar{\xi}_\delta(m) \cdot \Theta_F(\gamma m)\, dm = 0.$$

PROOF. Fix $m_0 \in M_0$. It would be sufficient to verify that

$$\int_{M_0 \cap K} \bar{\xi}_\delta(m_0 m) \cdot \Theta_F(\gamma m_0 m)\, dm = 0.$$

Put $f(k) = \Theta_F(\gamma m_0 k)$ and $g(m) = \xi_\delta(m_0 m)$ for $k \in K$ and $m \in M_0$. Then $f \in \mathcal{A}(F)$, and $g \in \mathcal{A}(\delta)$. Since $[F : \delta] = 0$, our assertion follows from Lemma 14.2.   □

For $i \geq 1$ let F_i be a finite subset of $\mathcal{E}(K)$ such that $\mathcal{E}(K)$ is the disjoint union of the F_i. Then, for $\beta \in C_c^\infty(M_0)$,

$$\theta(\beta) = \sum_i \int_{M_0} \beta(m) \cdot \Theta_{F_i}(\gamma m)\, dm.$$

In particular, for $\delta \in \mathcal{E}(M_0)$,

$$\theta(\bar{\xi}_\delta) = \sum_i \int_{M_0} \bar{\xi}_\delta(m) \cdot \Theta_{F_i}(\gamma m)\, dm.$$

LEMMA 18.3 (Lemma 36). *Fix $\delta \in \mathcal{E}(M_0)$. Then $\theta(\bar{\xi}_\delta) = 0$ unless we can choose i such that*

(1) δ *interacts with F_i and*
(2) $\Theta_{F_i}(\gamma m) \neq 0$ *for some $m \in M_0$.*

PROOF. This is obvious from Lemma 18.1 and Lemma 18.2.   □

19. First step of the reduction from G to M

Choose an $(M, \mathfrak{m})$-admissible M-domain $\mathfrak{m}_0$ in $\mathfrak{m} \cap \mathfrak{g}_0$ such that $\exp \mathfrak{m}_0 \subset U_M$. We shall use $\mathfrak{m}_0$ to define s-lattices in $\mathfrak{m}$.

Now fix s-lattices L and Λ in $\mathfrak{g}$ and $\mathfrak{m}$, respectively. Put $K = \exp L$, $K^{1/2} = \exp(\frac{1}{2}L)$, $M_0 = \exp \Lambda$, and $M_0^{1/2} = \exp(\frac{1}{2}\Lambda)$. We define $\mathcal{E}^{1/2}(K)$ and $\mathcal{E}^{1/2}(M_0)$ as usual. Let $p_{\mathfrak{m}}$ and $p_{\mathfrak{q}}$ denote the projections of $\mathfrak{g}$ on $\mathfrak{m}$ and $\mathfrak{q}$, respectively, corresponding to the direct sum $\mathfrak{g} = \mathfrak{m} + \mathfrak{q}$.

LEMMA 19.1. *Let Γ_1 and Γ_2 be two closed additive subgroups of $\mathfrak{g}$. If $\Gamma_1^* + \Gamma_2^*$ is closed, then*

$$(\Gamma_1 \cap \Gamma_2)^* = \Gamma_1^* + \Gamma_2^*.$$

PROOF. Since $\Gamma_1 \cap \Gamma_2 \subset \Gamma_i$, we have $\Gamma_i^* \subset (\Gamma_1 \cap \Gamma_2)^*$. So $\Gamma_1^* + \Gamma_2^* \subset (\Gamma_1 \cap \Gamma_2)^*$.

Note that $\Gamma_i^* \subset \Gamma_1^* + \Gamma_2^*$ which implies $(\Gamma_1^* + \Gamma_2^*)^* \subset \Gamma_1 \cap \Gamma_2$. So $(\Gamma_1 \cap \Gamma_2)^* \subset \Gamma_1^* + \Gamma_2^*$. $\square$

LEMMA 19.2 (Lemma 37). *Suppose $\underline{d} \in \mathcal{E}^{1/2}(K)$ and $\delta \in \mathcal{E}^{1/2}(M_0)$. Then $\underline{d}$ and δ interact if and only if $(p_{\mathfrak{m}}\mathcal{O}_{\underline{d}}) \cap \mathcal{O}_\delta \neq \emptyset$.*

PROOF. Let L^* be the lattice in $\mathfrak{g}$ dual to L. Let Λ^* be the lattice in $\mathfrak{m}$ dual to Λ. Then, for $X \in L$ and $u \in \Lambda$,

$$d(\underline{d}) \cdot \xi_{\underline{d}}(X) = d(\underline{d}) \cdot \xi_{\underline{d}}(\exp X) = \sum_{Y \in \mathcal{O}_{\underline{d}}/L^*} \chi\big(B(Y, X)\big)$$

and

$$d(\delta) \cdot \xi_\delta(u) = d(\delta) \cdot \xi_\delta(\exp u) = \sum_{Z \in \mathcal{O}_\delta/\Lambda^*} \chi\big(B(Z, u)\big).$$

Hence

$$d(\underline{d}) \cdot d(\delta) \int_{L \cap \Lambda} \xi_{\underline{d}}(u) \cdot \bar{\xi}_\delta(u) \, du = \sum_{Y \in \mathcal{O}_{\underline{d}}/L^*} \sum_{Z \in \mathcal{O}_\delta/\Lambda^*} \int_{L \cap \Lambda} \chi\big(B(Y - Z, u)\big) \, du.$$

But

$$\int_{L \cap \Lambda} \chi\big(B(Y - Z, u)\big) \, du = \begin{cases} \int_{L \cap \Lambda} du & \text{if } (Y - Z) \in (L \cap \Lambda)^*, \\ 0 & \text{otherwise.} \end{cases}$$

Now $(L \cap \Lambda)^* = L^* + \Lambda^* + \mathfrak{q}$. Hence $Y - Z \in (L \cap \Lambda)^* = L^* + \Lambda^* + \mathfrak{q}$ implies that

$$\mathcal{O}_{\underline{d}} \cap (\mathcal{O}_\delta + \mathfrak{q}) \neq \emptyset$$

or, equivalently,

$$(p_{\mathfrak{m}}\mathcal{O}_{\underline{d}}) \cap \mathcal{O}_\delta \neq \emptyset.$$

The converse is also true. $\square$

Put $K_0 = \exp L_0$ where L_0 is an s-lattice in $\mathfrak{g}$. By choosing L_0 sufficiently small we may assume that

(1) $K_0^{1/2} = \exp(\frac{1}{2}L_0) \subset GL(n, R)$,

(2) $\left|(\mathrm{Ad}(\exp\lambda)-1)X\right| \leq |\lambda||X|$ for $\lambda \in \frac{1}{2}L_0$ and $X \in \mathfrak{g}$, and

(3) Θ is (G, K_0)-admissible at γ.

Note that (3) implies that $K_0\gamma K_0 \subset U$.

Let $\mathbf{M}$ be the centralizer of γ in $\mathbf{G}$ and $\mathbf{M^0}$ the connected component of the identity in $\mathbf{M}$. Put $M^0 = \mathbf{M^0} \cap G$.

Fix an s lattice $\Lambda \subset \mathfrak{m}$ which is well adapted with respect to M^0. Put $M_0 = \exp\Lambda$. By choosing Λ sufficiently small, we may assume that

(1) $\Lambda \subset L_0$, $\frac{1}{2}\Lambda \subset \mathfrak{m}_0$, and $M_0 \subset U_M$,

(2) there exists a $c > 0$ such that $\left|(\mathrm{Ad}(\gamma m) - 1)Y\right| \geq c|Y|$ for $m \in M_0$ and $Y \in \mathfrak{q}$, and

(3) $\left|\frac{1}{2}\Lambda\right| < 1$ and $\left|(\mathrm{Ad}(m) - 1)Z\right| \leq \left|\frac{1}{2}\Lambda\right||Z|$ for $m \in M_0^{1/2}$ and $Z \in \mathfrak{m}$.

Choose an element $\varpi \in R$ of order one. For $\nu \in \mathbb{Z}$ and $\nu \geq 0$ define $L_\nu = \varpi^\nu L_0$, $K_\nu = \exp L_\nu$, and $K_\nu^{1/2} = \exp(\frac{1}{2}L_\nu)$. Note that for $m \in M_0$ and $\nu \geq 0$ we have $K_\nu\gamma m K_\nu \subset K_0\gamma K_0$. Let Φ_ν denote the set of all $\underline{d} \in \mathcal{E}^{1/2}(K_\nu)$ such that

(1) $\mathcal{O}_{\underline{d}} \cap \mathcal{N} \neq \emptyset$ and

(2) $\mathcal{O}_{\underline{d}}^{\gamma m} \cap \mathcal{O}_{\underline{d}} \neq \emptyset$ for some $m \in M_0$.

As usual let S be the set of all $X \in \mathfrak{g}$ such that $|X| = 1$ and let $\mathfrak{g}(1)$ denote the set of all $X \in \mathfrak{g}$ such that $|X| \leq 1$.

LEMMA 19.3 (Lemma 38). *Fix a neighborhood V of $\mathcal{N} \cap S \cap \mathfrak{m}$ in $S \cap \mathfrak{m}$. Then there exists an integer $\nu_0 \geq 0$ with the following property. Suppose $\nu \geq \nu_0$, $\underline{d} \in \Phi_\nu$, and $|X| \geq q^{2\nu}$ for some $X \in \mathcal{O}_{\underline{d}}$. Then*

$$p_{\mathfrak{m}}\mathcal{O}_{\underline{d}} \subset \Omega V.$$

We need some preparation before proving this lemma. For any endomorphism T of $\mathfrak{g}$, let $|T|$ denote the bound of T so that $|T| = \sup_{X \in \mathfrak{g}(1)} |T(X)|$. Put $q^r = \max(|p_{\mathfrak{m}}|, |p_{\mathfrak{q}}|)$. Then $r \geq 0$.

LEMMA 19.4. *We can choose a number $c_1 \geq \max(|\frac{1}{2}L_0|, |L_0^*|)$ with the following property. If $\nu \geq 0$ and $\underline{d} \in \Phi_\nu$, we can choose $X_0 \in \mathcal{O}_{\underline{d}}$ such that*

$$|p_{\mathfrak{q}}X_0| \leq c_1 \max(q^\nu, q^{-\nu}|X_0|)$$

and

$$|X_0 - X| \leq c_1 \max(q^\nu, q^{-\nu}|X_0|)$$

for all $X \in \mathcal{O}_{\underline{d}}$.

PROOF. Since $\underline{d} \in \Phi_\nu$, we can choose $m \in M_0$ such that $\mathcal{O}_{\underline{d}}^{\gamma m} \cap \mathcal{O}_{\underline{d}} \neq \emptyset$. Fix $X_0 \in \mathcal{O}_{\underline{d}}$ such that $\mathrm{Ad}(\gamma m)X_0 \in \mathcal{O}_{\underline{d}}$. Then we can choose $k \in K_\nu^{1/2}$ such that

$$\mathrm{Ad}(\gamma m)X_0 - \mathrm{Ad}(k)X_0 \in L_\nu^*.$$

Hence

$$\big(\mathrm{Ad}(\gamma m) - 1\big)X_0 - \big(\mathrm{Ad}(k) - 1\big)X_0 \in L_\nu^*.$$

Therefore

$$c\,|p_{\mathsf{q}}X_0| \leq \big|(\mathrm{Ad}(\gamma m) - 1)p_{\mathsf{q}}X_0\big| \leq \max\big(|p_{\mathsf{q}}L_\nu^*|, |p_{\mathsf{q}}(\mathrm{Ad}(k) - 1)X_0|\big).$$

Put $c_2 = c^{-1}|p_{\mathsf{q}}|$. Then

$$|p_{\mathsf{q}}X_0| \leq c_2 \max(q^\nu|L_0^*|, q^{-\nu}|X_0|\,|\tfrac{1}{2}L_0|).$$

Put $c_1 = \max(1, c_2) \cdot \max(|\tfrac{1}{2}L_0|, |L_0^*|)$. Then

$$|p_{\mathsf{q}}X_0| \leq c_1 \max(q^\nu, q^{-\nu}|X_0|).$$

Now fix $X \in \mathcal{O}_{\underline{d}}$. Then we can choose $k \in K_\nu^{1/2}$ and $\lambda \in L_\nu^*$ such that

$$X = \mathrm{Ad}(k)X_0 + \lambda.$$

Hence

$$X - X_0 = \big(\mathrm{Ad}(k) - 1\big)X_0 + \lambda,$$

and therefore

$$|X - X_0| \leq \max(|\tfrac{1}{2}L_\nu|\,|X_0|, |L_\nu^*|) \leq c_1 \max(q^\nu, q^{-\nu}|X_0|). \quad \square$$

COROLLARY 19.5. *Assume that $c_1 < q^{\nu-r}$ and $|X| \geq q^{2\nu}$ for some $X \in \mathcal{O}_{\underline{d}}$. Then, for all $X' \in \mathcal{O}_{\underline{d}}$,*

$$|p_{\mathsf{m}}X| = |X| = |X'|$$

and

$$|p_{\mathsf{m}}X - X'| \leq c_1\, q^{-\nu+r}\,|X|.$$

PROOF. Fix $X' \in \mathcal{O}_{\underline{d}}$. Then we can choose $k \in K_\nu^{1/2}$ and $\lambda \in L_\nu^*$ such that

$$X' = \mathrm{Ad}(k)X + \lambda.$$

Now

$$|\mathrm{Ad}(k)X| = |X|$$

since $K_\nu^{1/2} \subset K_0^{1/2} \subset GL(n, R)$. Moreover

$$|\lambda| \leq |L_\nu^*| \leq c_1\, q^\nu < q^{2\nu-r} \leq |X|.$$

Hence $|X'| = |X|$. In particular, $|X_0| = |X|$. But then

$$|p_{\mathsf{q}}X_0| \leq c_1 \max(q^\nu, q^{-\nu}|X_0|) < |X_0|.$$

Since $p_{\mathsf{m}}X_0 = X_0 - p_{\mathsf{q}}X_0$, we conclude that

$$|p_{\mathsf{m}}X_0| = |X_0|.$$

Also

$$|X_0 - X| \leq c_1 \max(q^\nu, q^{-\nu}|X_0|) \leq c_1\, q^{-\nu}|X_0| < q^{-r}\,|X_0|.$$

Hence, from the definition of r,

$$|p_{\mathsf{q}}(X_0 - X)| < |X_0|.$$

Now

$$p_{\mathsf{q}}X = p_{\mathsf{q}}(X - X_0) + p_{\mathsf{q}}(X_0).$$

Hence
$$|p_q X| < |X_0| = |X|.$$
This shows that $|p_m X| = |X - p_q X| = |X|$.

Finally
$$p_m X - X' = p_m(X - X_0) - (X' - X_0) - p_q(X_0).$$
We have seen above that
$$|X - X_0| \leq c_1 q^{-\nu}|X_0|.$$
This implies that $|p_m(X - X_0)| \leq c_1 q^{-\nu+r}|X_0|$ and $|X' - X_0| \leq c_1 q^{-\nu}|X_0|$ since $|X| = |X'|$. Also $|p_q X_0| \leq c_1 q^{-\nu}|X_0|$. Hence $|p_m X - X'| \leq c_1 q^{-\nu+r}|X_0| = c_1 q^{-\nu+r}|X|$. $\qquad\square$

Now we come to the proof of Lemma 19.3.

PROOF. Choose $\varepsilon \in (0,1)$ so small that
$$\bigl((\mathcal{N} \cap S) + \mathfrak{g}(\varepsilon)\bigr) \cap \mathfrak{m} \subset V.$$
Fix $\nu_0 \geq 0$ so that $c_1 q^{-\nu_0+r} \leq \varepsilon$.

Now fix $\nu \geq \nu_0$, $\underline{d} \in \Phi_\nu$, and suppose $X \in \mathcal{O}_{\underline{d}}$ such that $|X| \geq q^{2\nu}$. Choose $a \in \Omega$ such that $|a| = |X|$. Also choose $Y \in \mathcal{O}_{\underline{d}} \cap \mathcal{N}$. (Recall that $\mathcal{O}_{\underline{d}} \cap \mathcal{N} \neq \emptyset$ since $\underline{d} \in \Phi_\nu$.) Then by Corollary 19.5
$$|Y| = |X| = |a| \quad \text{and} \quad |p_m X - Y| \leq c_1 q^{-\nu+r}|X| \leq \varepsilon|a|.$$
Put $Y_0 = a^{-1}Y$ and $X_0 = a^{-1}X$. Then $Y_0 \in \mathcal{N} \cap S$, and
$$|p_m X_0 - Y_0| \leq \varepsilon.$$
Hence $p_m X_0 \in \bigl(\mathcal{N} \cap S + \mathfrak{g}(\varepsilon)\bigr) \cap \mathfrak{m} \subset V$. This proves that $p_m X \in \Omega V$. Moreover, it follows from Corollary 19.5 that $|X'| = |X| \geq q^{2\nu}$ for all $X' \in \mathcal{O}_{\underline{d}}$. By the above result, we have $p_m X' \in \Omega V$. This proves that $p_m \mathcal{O}_{\underline{d}} \subset \Omega V$. $\qquad\square$

20. Second step

Recall that the distribution θ is defined on U_M.

Put $\mathcal{O}_0 = L_0^*$. Then, by Theorem 17.1, $\mathcal{O}_0$ corresponds to 1_{K_0} regarded as an element of $\mathcal{E}^{1/2}(K_0)$. Fix $\nu \geq 0$ and $\underline{d} \in \mathcal{E}^{1/2}(K_\nu)$. Then we conclude from Corollary 17.2 that $\Theta_{\underline{d}} = 0$ on γK_0 unless
$$\mathcal{O}_{\underline{d}} \cap (\mathcal{O}_0)^G \neq \emptyset.$$

Define $\mathfrak{m}(t) = \mathfrak{m} \cap \mathfrak{g}(t)$ and recall that $q^r = \max(|p_m|, |p_q|)$. Since Λ is well adapted (with respect to M^0) the same holds for its dual Λ^* in $\mathfrak{m}$. Hence, by Theorem 13.1, we can choose a neighborhood V of $\mathcal{N} \cap S \cap \mathfrak{m}$ in $S \cap \mathfrak{m}$ and a number $t_0 > 0$ such that $C(V, t_0, \Lambda^*)$ holds. Moreover, we may assume that $V = (\mathcal{N} \cap S \cap \mathfrak{m}) + \mathfrak{m}(\varepsilon)$ for some $\varepsilon \in (0,1)$.

For $\delta \in \mathcal{E}^{1/2}(M_0)$, put
$$\theta(\delta) = \theta(\bar\xi_{\delta_0})$$

where δ_0 is any element of $\mathcal{E}(M_0)$ lying in the orbit δ. Since θ is M-invariant, this definition is legitimate. For $\nu \geq 0$ let F_ν denote the set of all $\delta \in \mathcal{E}^{1/2}(M_0)$ such that $\mathcal{O}_\delta \subset \mathfrak{m}(q^{2\nu+r})$. F_ν is a finite set.

Fix $\nu_0 \geq 0$ so large that

(1) $\mathcal{O}_0^G \subset \mathcal{N} + L_{\nu_0}^*$,
(2) $q^{2\nu_0} > \max(t_0, |\Lambda^*||\frac{1}{2}\Lambda|^{-1}\varepsilon^{-1}, (c_1 q^r)^2)$, and
(3) Lemma 19.3 is satisfied.

Here c_1 is the constant of Lemma 19.4.

LEMMA 20.1. *Fix* $\nu \geq \nu_0$, $\delta \in \mathcal{E}^{1/2}(M_0)$ *and suppose* $\delta \notin F_\nu$. *Then* $\mathcal{O}_\delta \cap$ $\mathfrak{m}(q^{2\nu+r}) = \emptyset$.

PROOF. Since $\delta \notin F_\nu$, we can choose $Z \in \mathcal{O}_\delta$ such that $|Z| > q^{2\nu+r}$. If $Z' \in \mathcal{O}_\delta$, we can choose $m \in M_0^{1/2}$ and $\lambda^* \in \Lambda^*$ such that

$$Z' = \mathrm{Ad}(m)Z + \lambda^*.$$

Then

$$Z' - Z = \big(\mathrm{Ad}(m) - 1\big)Z + \lambda^*,$$

and therefore

$$|Z' - Z| \leq \max(|\tfrac{1}{2}\Lambda||Z|, |\Lambda^*|).$$

But

$$|Z| > q^{2\nu+r} \geq q^{2\nu_0} \geq |\Lambda^*||\tfrac{1}{2}\Lambda|^{-1},$$

and $|\frac{1}{2}\Lambda| < 1$. Hence

$$|Z' - Z| \leq |\tfrac{1}{2}\Lambda||Z| < |Z|,$$

and therefore $|Z'| = |Z| > q^{2\nu+r}$. This shows that $\mathcal{O}_\delta \cap \mathfrak{m}(q^{2\nu+r}) = \emptyset$. □

LEMMA 20.2 (Lemma 39). *Fix* $\nu \geq \nu_0$. *If* $\delta \in \mathcal{E}^{1/2}(M_0)$, $\delta \notin F_\nu$ *and* $\theta(\delta) \neq 0$, *we can conclude that*

$$\mathcal{O}_\delta \subset \Omega V$$

and

$$|Z| > q^{2\nu} \geq t_0$$

for all $Z \in \mathcal{O}_\delta$.

PROOF. Since $\theta(\delta) \neq 0$, it follows from Lemma 18.3 and Lemma 19.2 that we can choose $\underline{d} \in \mathcal{E}^{1/2}(K_\nu)$ such that

(20.1) $$(p_{\mathfrak{m}}\mathcal{O}_{\underline{d}}) \cap \mathcal{O}_\delta \neq \emptyset$$

and

(20.2) $$\Theta_{\underline{d}}(\gamma m) \neq 0$$

for some $m \in M_0$.

As noted at the beginning of this section, $\mathcal{O}_{\underline{d}} \cap \mathcal{O}_0^G \neq \emptyset$. But $\mathcal{O}_0^G \subset \mathcal{N} + L_\nu^*$, and therefore

$$(20.3) \qquad\qquad\qquad\qquad \mathcal{O}_{\underline{d}} \cap \mathcal{N} \neq \emptyset.$$

By equation (20.2) we may fix an $m \in M_0$ such that $\Theta_{\underline{d}}(\gamma m) \neq 0$. Note that $K_\nu \gamma m K_\nu \subset K_0 \gamma K_0 \subset U$, and the function

$$(k_1, k_2) \mapsto \Theta_{\underline{d}}(k_1 \gamma m k_2)$$

on $K_\nu \times K_\nu$ lies in $\mathcal{A}(\underline{d}) \otimes \mathcal{A}(\underline{d})$. Hence, we conclude from Corollary 14.3 and Corollary 17.2 that

$$(20.4) \qquad\qquad\qquad\qquad \mathcal{O}_{\underline{d}} \cap \mathcal{O}_{\underline{d}}{}^{\gamma m} \neq \emptyset$$

for some $m \in M_0$. Equation (20.3) and equation (20.4) show that $\underline{d} \in \Phi_\nu$.

By equation (20.1) we can choose $X \in \mathcal{O}_{\underline{d}}$ such that $Z = p_{\mathrm{m}} X \in \mathcal{O}_\delta$. We claim that $|X| > q^{2\nu}$. Suppose not. If $|X| \leq q^{2\nu}$, then $|Z| = |p_{\mathrm{m}} X| \leq q^{2\nu+r}$. But since $\delta \notin F_\nu$, we know from Lemma 20.1 that $\mathcal{O}_\delta \cap \mathrm{m}(q^{2\nu+r}) = \emptyset$, and therefore this is impossible. This shows that $|X| > q^{2\nu}$, and so, from Corollary 19.5, we have

$$|Z| = |p_{\mathrm{m}} X| = |X| > q^{2\nu}.$$

Recall that $V = (\mathcal{N} \cap S \cap \mathrm{m}) + \mathrm{m}(\varepsilon)$ for some $\varepsilon \in (0, 1)$. Since $M_0^{1/2} \subset K_0^{1/2} \subset GL(n, R)$, we have $V^m = V$ for $m \in M_0^{1/2}$. Fix $a \in \Omega$ such that

$$|a| = |X| = |Z| > q^{2\nu}.$$

Then, by Lemma 19.3, $a^{-1} Z \in V$.

Now let $Z' \in \mathcal{O}_\delta$. Then we can choose $m \in M_0^{1/2}$ and $\lambda^* \in \Lambda^*$ such that

$$Z' = Z^m + \lambda^*.$$

Then $a^{-1} Z^m \in V^m = V$, and

$$|Z^m| = |Z| = |a| > q^{2\nu}.$$

Moreover

$$|a^{-1} \lambda^*| \leq q^{-2\nu} |\Lambda^*| \leq \varepsilon.$$

This shows that $a^{-1} Z' \in V + \mathrm{m}(\varepsilon) = V$, and hence $\mathcal{O}_\delta \subset aV \subset \Omega V$. $\qquad \square$

21. Completion of the proof

For $f \in C_c^\infty(\mathrm{m})$, define $f_0 \in C_c^\infty(U_M)$ as follows.

$$f_0(u) = \begin{cases} f(Z) & \text{if } u = \exp Z \text{ and } Z \in \mathrm{m}_0, \\ 0 & \text{otherwise.} \end{cases}$$

Put

$$\theta_0(f) = \theta(f_0).$$

Then θ_0 is an M-invariant distribution on m and $\mathrm{Supp}(\theta_0) \subset \mathrm{m}_0$.

LEMMA 21.1 (Lemma 40). *For $Z \in \mathfrak{m}$ let f_Z denote the characteristic function of $Z + \Lambda^*$ on $\mathfrak{m}$. Then if $\delta \in \mathcal{E}^{1/2}(M_0)$ and $Z \in \mathcal{O}_\delta$,*

$$\theta(\delta) = \nu(\Lambda^*)^{-1} \cdot d(\delta) \cdot \widehat{\theta}_0(f_{-Z})$$

where $\nu(\Lambda^*) = \int_{\Lambda^*} du$.

PROOF. We regard ξ_δ as a function on $\mathfrak{m}$ as follows.

$$\xi_\delta(Z) = \begin{cases} \xi_\delta(\exp Z) & \text{for } Z \in \Lambda, \\ 0 & \text{otherwise.} \end{cases}$$

Let α_δ be the characteristic function of $\mathcal{O}_\delta$. Then

$$\widehat{\alpha_\delta}(Z) = \int_{\mathfrak{g}} \chi(B(Z, u)) \cdot \alpha_\delta(u) \, du.$$

Clearly $\widehat{\alpha_\delta} = 0$ outside Λ. On the other hand, if $Z \in \Lambda$,

$$\widehat{\alpha_\delta}(Z) = \nu(\Lambda^*) \sum_{u \in \mathcal{O}_\delta/\Lambda^*} \chi(B(Z, u)).$$

It follows from Lemma 17.1 that

$$\widehat{\alpha_\delta} = \nu(\Lambda^*) \cdot d(\delta) \cdot \xi_\delta.$$

For any $g \in C_c^\infty(\mathfrak{m})$ define $\breve{g}$ by $\breve{g}(Z) = g(-Z)$. Then

$$\theta(\bar\xi_\delta) = \theta_0(\breve\xi_\delta) = \frac{\widehat{\theta}_0(\breve\alpha_\delta)}{\nu(\Lambda^*) \cdot d(\delta)}.$$

Now

$$\breve\alpha_\delta = \sum_{Z \in \mathcal{O}_\delta/\Lambda^*} f_{-Z}.$$

Since θ_0 is M-invariant and $\mathcal{O}_\delta$ is a single $M_0^{1/2}$-orbit in $\mathfrak{m}/\Lambda^*$, we conclude that

$$\widehat{\theta}_0(\breve\alpha_\delta) = [\mathcal{O}_\delta/\Lambda^*] \cdot \widehat{\theta}_0(f_{-Z})$$

where Z is any element of $\mathcal{O}_\delta$. But, by Theorem 17.1,

$$[\mathcal{O}_\delta/\Lambda^*] = d(\delta) \cdot \xi_\delta(1) = d(\delta)^2 \cdot N(\delta)$$

where $N(\delta)$ is the number of elements of $\mathcal{E}(M_0)$ lying on the orbit δ. Hence

$$\theta(\bar\xi_\delta) = \frac{d(\delta) \cdot N(\delta) \cdot \widehat{\theta}_0(f_{-Z})}{\nu(\Lambda^*)}.$$

But it is obvious that $\theta(\delta) \cdot N(\delta) = \theta(\bar\xi_\delta)$. Hence

$$\theta(\delta) = \frac{d(\delta) \cdot \widehat{\theta}_0(f_{-Z})}{\nu(\Lambda^*)}. \quad \square$$

Fix ν as in Lemma 20.2.

LEMMA 21.2 (Lemma 41). *Suppose Z is an element of $\mathfrak{m}$ such that $|Z| > q^{2\nu+r}$ and $Z \notin \Omega V$. Then*

$$\widehat{\theta}_0(f_Z) = 0.$$

PROOF. There is a unique $\delta \in \mathcal{E}^{1/2}(M_0)$ such that $-Z \in \mathcal{O}_\delta$. Since $|Z| > q^{2\nu+r}$, it is clear that $\delta \notin F_\nu$. We claim that $\theta(\delta) = 0$. If this were not so, then, by Lemma 20.2,

$$-Z \in \mathcal{O}_\delta \subset \Omega V,$$

and this contradicts our hypothesis. Hence

$$0 = \theta(\delta) = \nu(\Lambda^*)^{-1} \cdot d(\delta) \cdot \widehat{\theta}_0(f_Z)$$

from Lemma 21.1. This proves that $\widehat{\theta}_0(f_Z) = 0$.   □

Lemma 21.2 shows that $\widehat{\theta}_0 \in J(V, \infty, \Lambda^*)$ (in the notation of §11.1 applied to $(M^0, \mathfrak{m})$ in place of $(G, \mathfrak{g})$). We conclude from Corollary 11.4 that $j_{\Lambda^*} \widehat{\theta}_0 \in j_{\Lambda^*} J_0$. Hence, we can choose $\tau \in J_0$ such that $j_{\Lambda^*} \widehat{\theta}_0 = j_{\Lambda^*} \check{\tau}$. This means that $\theta_0 = \hat{\tau}$ on Λ.

It follows from Theorem 4.4 that θ_0 is a locally summable function on Λ. Hence, θ is a locally summable function on M_0. Since $(\gamma M_0)^G$ is a neighborhood of γ in U, it follows from Theorem 11 and its Corollary in [21] that Θ is a locally summable function around γ, whence Theorem 16.1.

In fact,

$$\Theta(\gamma \exp Y) = \theta_0(Y)$$

for Y sufficiently near zero in $\mathfrak{m}$. Since $\theta_0 = \hat{\tau}$ near zero, by Theorem 5.11 there exist complex numbers c_ξ such that

$$\theta_0(Y) = \sum c_\xi \cdot \widehat{\nu}_\xi(Y)$$

for Y near zero in $\mathfrak{m}$. Here, as usual, ξ runs over all nilpotent M-orbits in $\mathfrak{m}$, ν_ξ is the M-invariant measure on $\mathfrak{m}$ corresponding to ξ, and $\widehat{\nu}_\xi$ is the Fourier transform of ν_ξ on $\mathfrak{m}$. Whence Theorem 16.2.

Let $\mathcal{O}$ be a G-orbit in G. Then the closure of $\mathcal{O}$ contains a semisimple element γ. Theorem 16.3 is a consequence of this fact.

22. Formal degree of a supercuspidal representation

Fix a Haar measure dx^* on G/Z where Z is the maximal split torus lying in the center of G. For any square-integrable representation π of G, let $d(\pi)$ denote the formal degree of π [20, §3] defined by means of dx^*.

A Cartan subgroup Γ of G is called elliptic if Γ/Z is compact. Define D as in §17, and let G' be the set of all $x \in G$ where $D(x) \neq 0$. Fix an elliptic Cartan subgroup Γ of G, and put $\Gamma' = \Gamma \cap G'$. If $f \in C^\infty(G)$ such that $\mathrm{Supp}(f)$ is compact mod Z, we define, for $\gamma \in \Gamma'$,

$$F_f(\gamma) = |D(\gamma)|^{1/2} \int_{G/Z} f(x\gamma x^{-1}) \, dx^*.$$

Then F_f is a locally constant function on Γ' (see, e.g., [22, Theorem 3]).

Let π be an irreducible, unitary, supercuspidal representation of G [20, §6]. Define $\mathcal{A}(\pi)$ as usual [20, §3] and let Θ_π denote the character of π. We know from Theorem 16.3 that Θ_π is a function.

LEMMA 22.1 (Lemma 42). *For $f \in \mathcal{A}(\pi)$ and $\gamma \in \Gamma'$,*

$$F_f(\gamma) = d(\pi)^{-1} \cdot f(1) \cdot |D(\gamma)|^{1/2} \cdot \Theta_\pi(\gamma).$$

This follows from the Schur orthogonality relations [20, §3] by standard arguments.

Let $\mathfrak{g}_0$ be a $(G, \mathfrak{g})$-admissible G-domain and $G_0 = \exp(\mathfrak{g}_0)$. Let dX denote a Haar measure on the additive group of $\mathfrak{g}$. Then it follows that the Haar measure dx on G can be normalized so as to correspond to dX under the exponential mapping. Let θ_π denote the distribution on $\mathfrak{g}_0$ given by

$$\theta_\pi(a_0) = \Theta_\pi(a)$$

where $a \in C_c^\infty(G_0)$ and $a_0 \in C_c^\infty(\mathfrak{g}_0)$ is defined by $a_0(X) = a(\exp X)$ for $X \in \mathfrak{g}_0$. Then it is obvious that

$$\theta_\pi(a_0) = \int_{\mathfrak{g}_0} a_0(X) \cdot \theta_\pi(X) \, dX$$

where

$$\theta_\pi(X) = \Theta_\pi(\exp X)$$

for $X \in \mathfrak{g}_0$.

Fix $f \in \mathcal{A}(\pi)$ and define the function g on $\mathfrak{g}$ by

$$g(X) = \begin{cases} f(\exp X) & \text{if } X \in \mathfrak{g}_0, \\ 0 & \text{otherwise.} \end{cases}$$

Then it follows from condition (2) on $\mathfrak{g}_0$ (page 58) that $g \in C_c^\infty(\mathfrak{g}_0)$. Let $\mathfrak{h}$ be the Lie algebra of Γ. It is clear that

$$F_f(\exp H) = \phi_g^{\mathfrak{h}}(H)$$

for $H \in \mathfrak{h}' \cap \mathfrak{g}_0$. Hence we obtain the following result from Lemma 22.1 and Theorem 8.1.

LEMMA 22.2 (Lemma 43). *There exists a neighborhood ω of zero in $\mathfrak{h} \cap \mathfrak{g}_0$ such that*

$$|\eta(H)|^{1/2} \cdot \theta_\pi(H) \cdot g(0) = d(\pi) \sum_{\mathcal{O} \in \mathcal{O}(0)} \mu_\mathcal{O}(g) \cdot \Gamma_\mathcal{O}^{\mathfrak{h}}(H)$$

for all $H \in \omega \cap \mathfrak{h}'$.

On the other hand, we know from Theorem 16.2 that

$$\theta_\pi = \sum_{\mathcal{O}} c_\mathcal{O}(\pi) \cdot \widehat{\mu_\mathcal{O}}$$

around zero. Let 0 denote the orbit $\mathcal{O} = \{0\}$. We may assume that $\mu_0(h) = h(0)$ for all $h \in \mathcal{D}$. Hence $\widehat{\mu_0} = 1$ and Lemma 22.2 can be expressed in the notation of §9 as follows.

$$\left[|\eta(H)|^{1/2} \cdot c_0(\pi) + \sum_{\mathcal{O} \neq 0} c_{\mathcal{O}}(\pi) \cdot \Psi_{\mathcal{O}}(H) \right] \cdot g(0)$$

$$= d(\pi) \cdot \left[\Gamma_0^{\mathfrak{h}}(H) \cdot g(0) + \sum_{\mathcal{O} \neq 0} \mu_{\mathcal{O}}(g) \cdot \Gamma_{\mathcal{O}}^{\mathfrak{h}}(H) \right]$$

for $H \in \omega \cap \mathfrak{h}'$. Since 0 is the only nilpotent G-orbit of dimension zero, we conclude from Lemma 9.2 and part (1) of Theorem 8.1 that

$$c_0(\pi) = c\, d(\pi)$$

where c is defined as in Theorem 9.6. This proves the following theorem.

THEOREM 22.3 (Theorem 6). *There exists a real number $c \neq 0$ such that*

$$c_0(\pi) = c\, d(\pi)$$

for every irreducible, unitary, supercuspidal representation π.

Put $K = \exp L$ where L is an s-lattice in $\mathfrak{g}$, and let $m_\pi(L)$ denote the multiplicity of the trivial representation of K in π.

LEMMA 22.4 (Lemma 44). *If α_L is the characteristic function of L and*

$$\nu(L) = \int_L dX,$$

then $m_\pi(L) = \nu(L)^{-1} \cdot \theta_\pi(\alpha_L)$.

PROOF. Let dk denote the normalized measure on K and α_K the characteristic function of K. It is obvious that

$$\int_K dx = \int_L dX = \nu(L),$$

and therefore $dk = \nu(L)^{-1} dx$. Now $m_\pi(L) = \operatorname{tr} E$ where

$$E = \int_K \pi(k)\, dk = \nu(L)^{-1} \int_G \alpha_K(x) \cdot \pi(x)\, dx.$$

Hence $m_\pi(L) = \nu(L)^{-1} \cdot \Theta_\pi(\alpha_K) = \nu(L)^{-1} \cdot \theta_\pi(\alpha_L)$. $\qquad\square$

Now assume that L is so small that

$$\theta_\pi = \sum_{\mathcal{O}} c_{\mathcal{O}}(\pi) \cdot \widehat{\mu_{\mathcal{O}}}$$

on L. Then

$$\theta_\pi(\alpha_L) = \sum_{\mathcal{O}} c_{\mathcal{O}}(\pi) \cdot \mu_{\mathcal{O}}\big((\alpha_L)^{\widehat{}}\big).$$

But

$$(\alpha_L)^{\widehat{}} = \nu(L)\alpha_{L^*}.$$

where α_{L^*} is the characteristic function of the dual lattice L^*. For $t \in R'$ replace L by t^2L, and observe that $(t^2L)^* = t^{-2}L^*$. Note that

$$\mu_{\mathcal{O}}(\alpha_{t^{-2}L^*}) = |t|^{-d(\mathcal{O})}\mu_{\mathcal{O}}(\alpha_{L^*})$$

from Lemma 3.2. This gives the following result.

LEMMA 22.5 (Lemma 45). *If L is sufficiently small,*

$$m_\pi(\varpi^{2j}L) = \sum_{\mathcal{O}} c_{\mathcal{O}}(\pi) \cdot q^{jd(\mathcal{O})} \cdot \mu_{\mathcal{O}}(\alpha_{L^*})$$

for all $j \geq 0$.

Let $d_0 = 0 < d_1 < \cdots < d_r$ be all the integers d such that $d = d(\mathcal{O})$ for some $\mathcal{O} \in \mathcal{O}(0)$. Put

$$c(d) = \sum_{d(\mathcal{O})=d} c_{\mathcal{O}}(\pi) \cdot \mu_{\mathcal{O}}(\alpha_{L^*}).$$

Then we conclude from Lemma 22.5 that

$$\sum_{0 \leq i \leq r} c(d_i) \cdot q^{jd_i} = m_\pi(\varpi^{2j}L) \in \mathbb{Z}$$

for $0 \leq j \leq r$. Put $x_i = q^{d_i}$. Then

$$\sum_{0 \leq i \leq r} c(d_i) \cdot x_i^j \in \mathbb{Z}$$

for $0 \leq j \leq r$ and

$$\det(x_i^j) = \prod_{0 \leq i < j \leq r} (x_j - x_i) = Q \neq 0.$$

Hence $Q\,c(d_i) \in \mathbb{Z}$. But

$$c(0) = c_0(\pi) \cdot \mu_0(\alpha_{L^*}) = c_0(\pi) = c\,d(\pi)$$

from Theorem 22.3, and therefore we conclude that $Q\,c\,d(\pi) \in \mathbb{Z}$. This proves the following theorem.

THEOREM 22.6 (Theorem 7). *It is possible to normalize the Haar measure on G/Z in such a way that $d(\pi)$ is an integer for every irreducible, unitary, supercuspidal representation π.*

Bibliography

[1] M. Assem, *The Fourier transform and some character formulae for p-adic* SL_ℓ, ℓ *a prime*, Amer. J. Math. **116** (1994), no. 6, pp. 1433–1467.

[2] D. Barbasch and A. Moy, *Local character expansions*, Ann. Sci. Éc. Norm. Sup. (4) **30** (1997), no. 5, pp. 553–567.

[3] E. Bosman, *Harmonic analysis on p-adic symmetric spaces*, Ph.D. thesis, Rijkuniversiteit te Leiden, 1992.

[4] L. Clozel, *Characters of non-connected, reductive p-adic groups*, Canad. J. Math., **39** (1987), pp. 149–167.

[5] F. Courtès, *Sur le transfert des intégrales orbitales pour les groupes linéaires (cas p-adique)*, Mém. Soc. Math. Fr. (N.S.) No. 69 (1997), vi+140 pp.

[6] C. Cunningham, *The characters of depth zero representations of* $Sp_4(F)$, Ph.D. thesis, University of Toronto, 1997.

[7] S. DeBacker, *On supercuspidal characters of* GL_ℓ, ℓ *a prime*, Ph.D. thesis, The University of Chicago, 1997.

[8] S. DeBacker and P. J. Sally, Jr., *Germs, characters, and the Fourier transforms of nilpotent orbits*, to appear.

[9] Y. Flicker, *Orbital integrals on symmetric spaces and spherical characters*, J. Algebra **184** (1996), no. 2, pp. 705–754.

[10] Y. Flicker and D. Kazhdan, *Metaplectic correspondence*, Inst. Hautes Études Sci. Publ. Math. No. 64 (1986), pp. 53–110.

[11] J. Hakim, *Admissible distributions on p-adic symmetric spaces*, J. Reine Angew. Math. **455** (1994), pp. 1–19.

[12] _____, *Howe/Kirillov theory for p-adic symmetric spaces*, Proc. Amer. Math. Soc. **121** (1994), no. 4, pp. 1299–1305.

[13] T. Hales, *Hyperelliptic curves and harmonic analysis (why harmonic analysis on reductive p-adic groups is not elementary)*, Representation theory and analysis on homogeneous spaces (New Brunswick, NJ, 1993), Contemp. Math., vol. 177, Amer. Math. Soc., Providence, RI, 1994, pp. 137–169.

[14] _____, *Shalika germs on* GSp(4), Orbites unipotentes et représentations, II. Astérisque No. 171-172 (1989), pp. 195–256.

[15] _____, *The subregular germ of orbital integrals*, Mem. Amer. Math. Soc. **99** (1992), no. 476.

[16] _____, *The twisted endoscopy of* GL(4) *and* GL(5): *transfer of Shalika germs*, Duke Math. J. **76** (1994), no. 2, pp. 595–632.

[17] Harish-Chandra, *Admissible invariant distributions on reductive p-adic groups*, Lie theories and their applications (Proc. Ann. Sem. Canad. Math. Congr., Queen's Univ., Kingston, Ont., 1977), Queen's Papers in Pure Appl. Math., No. 48, Queen's Univ., Kingston, Ont., 1978, pp. 281–347.

[18] _____, *The characters of reductive p-adic groups*, Contributions to algebra (collection of papers dedicated to Ellis Kolchin), Academic Press, New York, 1977, pp. 175–182.

[19] _____, *Fourier transforms on a semisimple Lie algebra I*, Amer. J. Math., **79** (1957), pp. 193–257.

[20] _____, *Harmonic analysis on reductive p-adic groups*, Harmonic analysis on homogeneous spaces (Proc. Sympos. Pure Math., Vol. XXVI, Williams Coll., Williamstown, Mass., 1972), Amer. Math. Soc., Providence, R.I., 1973, pp. 167–192.

[21] _____, (notes by G. van Dijk), *Harmonic analysis on reductive p-adic groups*, Lecture Notes in Mathematics, vol. 162, Springer, 1970.

[22] _____, *A submersion principle and its applications*, in *Geometry and Analysis – Papers Dedicated to the Memory of V. K. Patodi*, Springer-Verlag, 1981, pp. 95–102.

[23] R. Howe, *The Fourier transform and germs of characters (Case of Gl_n over a p-adic field)*, Math. Ann. **208** (1974), pp. 305–322.

[24] _____, *Kirillov theory for compact p-adic groups*, Pacific Journal of Mathematics, **73** (1977), pp. 365–381.

[25] _____, *Some qualitative results on the representation theory of GL_n over a p-adic field*, Pacific Journal of Mathematics, **73** (1977), pp. 479–538.

[26] _____, *Two conjectures about reductive p-adic groups*, Harmonic analysis on homogeneous spaces (Proc. Sympos. Pure Math., Vol. XXVI, Williams Coll., Williamstown, Mass., 1972), Amer. Math. Soc., Providence, R.I., 1973, pp. 377–380.

[27] R. Huntsinger, *Some aspects of invariant harmonic analysis on the Lie algebra of a reductive p-adic group*, Ph.D. thesis, The University of Chicago, 1997.

[28] _____, *Vanishing of the leading term in Harish-Chandra's local character expansion*, Proc. Amer. Math. Soc. **124** (1996), no. 7, pp. 2229–2234.

[29] H. Jacquet, *Sur les représentations des groupes réductifs p-adiques*, C. R. Acad. Sci. Paris Sér. A-B **280** (1975), pp. A1271–A1272.

[30] D. Kazhdan, *Cuspidal geometry of p-adic groups*, J. Analyse Math. **47** (1986), pp. 1–36.

[31] _____, *On lifting*, Lie group representations, II (College Park, Md., 1982/1983), Lecture Notes in Math., vol. 1041, Springer, Berlin-New York, 1984, pp. 209–249.

[32] D. Kazhdan and G. Savin, *The smallest representation of simply laced groups*, Festschrift in honor of I. I. Piatetski-Shapiro on the occasion of his sixtieth birthday, Part I (Ramat Aviv, 1989), pp. 209–223, Israel Math. Conf. Proc., 2, Weizmann, Jerusalem, 1990.

[33] Y. Kim, *An example of subregular germs for 4×4 symplectic groups*, Honam Math. J. **15** (1993), no. 1, pp. 47–53.

[34] _____, *On whole regular germs for p-adic $Sp_4(F)$*, J. Korean Math. Soc. **28** (1991), no. 2, pp. 207–213.

[35] _____, *Regular germs for p-adic Sp(4)*, Canad. J. Math. **41** (1989), no. 4, pp. 626–641.

[36] Y. Kim and K. So, *Some subregular germs for p-adic $Sp_4(F)$*, Internat. J. Math. Math. Sci. **18** (1995), no. 1, pp. 37–47.

[37] R. Kottwitz, *Harmonic analysis on semisimple p-adic Lie algebras*, Proceedings of the International Congress of Mathematicians, Vol. II (Berlin, 1998). Doc. Math. 1998, Extra Vol. II, pp. 553–562 (electronic).

[38] R. P. Langlands and D. Shelstad, *Orbital integrals on forms of SL(3). II.*, Canad. J. Math. **41** (1989), no. 3, pp. 480–507.

[39] B. Lemaire, *Intégrabilité locale des caractères-distributions de $GL_N(F)$ où F est un corps local non-archimédien de caractéristique quelconque*, Compositio Math. **100** (1996), no. 1, pp. 41–75.

[40] C. Mœglin, *Front d'onde des représentations des groupes classiques p-adiques*, Amer. J. Math. **118** (1996), no. 6, pp. 1313–1346.

[41] C. Mœglin and J.-L. Waldspurger, *Modèles de Whittaker dégénérés pour des groupes p-adiques*, Math. Z. **196** (1987), no. 3, pp. 427–452.

[42] A. Moy and G. Prasad, *Jacquet functors and unrefined minimal K-types*, Comment. Math. Helvetici **71** (1996), pp. 98–121.

[43] ———, *Unrefined minimal K-types for p-adic groups*, Inv. Math. **116** (1994), pp. 393–408.

[44] F. Murnaghan, *Asymptotic behaviour of supercuspidal characters*, Representation theory of groups and algebras, Contemp. Math., **145**, Amer. Math. Soc., Providence, RI, 1993, pp. 155–162.

[45] ———, *Asymptotic behaviour of supercuspidal characters of p-adic* GL_3 *and* GL_4: *the generic unramified case*, Pacific J. Math. **148** (1991), no. 1, pp. 107–130.

[46] ———, *Asymptotic behaviour of supercuspidal characters of p-adic* $GSp(4)$, Compositio Math. **80** (1991), no. 1, pp. 15–54.

[47] ———, *Characters of supercuspidal representations of classical groups*, Ann. Sci. Éc. Norm. Sup. (4) **29** (1996), no. 1, pp. 49–105.

[48] ———, *Characters of supercuspidal representations of* $SL(n)$, Pacific J. Math. **170** (1995), no. 1, pp. 217–235.

[49] ———, *Germs of characters of admissible representations*, to appear.

[50] ———, *Local character expansions and Shalika germs for* $GL(n)$, Math. Ann. **304** (1996), no. 3, pp. 423–455.

[51] ———, *Local character expansions for supercuspidal representations of* $U(3)$, Canad. J. Math. **47** (1995), no. 3, pp. 606–640.

[52] F. Murnaghan and J. Repka, *Vanishing of coefficients in overlapping germ expansions for p-adic* $GL(n)$, Proc. Amer. Math. Soc. **111** (1991), no. 4, pp. 1183–1193.

[53] C. Rader and S. Rallis, *Spherical characters on p-adic symmetric spaces*, Amer. J. Math. **118** (1996), no. 1, pp. 91–178.

[54] C. Rader and A. Silberger, *Some consequences of Harish-Chandra's submersion principle*, Proc. Amer. Math. Soc. **118** (1993), no. 4, pp. 1271–1279.

[55] R. Ranga Rao, *Orbital integrals in reductive groups*, Ann. of Math. **96** (1972), pp. 505–510.

[56] J. Repka, *Germs associated to regular unipotent classes in p-adic* $SL(n)$, Canad. Math. Bull. **28** (1985), no. 3, pp. 257–266.

[57] ———, *Shalika's germs for p-adic* $GL(n)$. *I. The leading term*, Pacific J. Math. **113** (1984), no. 1, pp. 165–172.

[58] ———, *Shalika's germs for p-adic* $GL(n)$. *II. The subregular term*, Pacific J. Math. **113** (1984), no. 1, pp. 173–182.

[59] F. Rodier, *Intégrabilité locale des caractères du groupe $GL(n,k)$ où k est un local corps de caratéristique positive*, Duke Math. J., **52** (1985), pp. 771–792.

[60] ———, *Modèle de Whittaker et caractères de représentations*, Non-commutative harmonic analysis (Actes Colloq., Marseille-Luminy, 1974), pp. 151–171, Lecutre Notes in Math., 466, Springer, Berlin, 1975.

[61] J. Rogawski, *An application of the building to orbital integrals*, Compositio Math. **42** (1980/81), no. 3, pp. 417–423.

[62] ———, *Some remarks on Shalika germs*, The Selberg trace formula and related topics (Brunswick, Maine, 1984), Contemp. Math., vol. 53, Amer. Math. Soc., Providence, R.I., 1986, pp. 387–391.

[63] K. Rumelhart, *Minimal representations of exceptional p-adic groups*, Represent. Theory **1** (1997), pp. 133–181 (electronic).

[64] P. J. Sally, Jr. and J. Shalika, *The Fourier transform of orbital integrals on* SL_2 *over a p-adic field*, Lie group representations, II (College Park, Md., 1982/1983), Lecture Notes in Math., **1041**, Springer, Berlin-New York, 1984, pp. 303–340.

[65] G. Savin, *Dual pair* $G_{\mathfrak{J}} \times PGL_2$ *[where]* $G_{\mathfrak{J}}$ *is the automorphism group of the Jordan algebra* $\mathfrak{J}$, Invent. Math. **118** (1994), no. 1, pp. 141–160.

[66] J. Shalika, *A theorem on semi-simple p-adic groups*, Ann. of Math. **95** (1972), pp. 226–242.

[67] D. Shelstad, *A formula for regular unipotent germs*, Orbites unipotentes et représentations, II. Astérisque No. 171-172 (1989), pp. 275–277.

[68] A. Silberger, *Introduction to Harmonic Analysis on Reductive p-adic Groups*, Princeton University Press, 1979.

[69] P. Torasso, *Méthode des orbites de Kirillov-Duflo et représentations minimales des groupes simples sur un corps local de caractéristique nulle*, Duke Math. J. **90** (1997), no. 2, pp. 261–377.

[70] M.-F. Vignéras, *Caractère d'une représentation modulaire d'une groupe p-adique*, 1998, preprint.

[71] J.-L. Waldspurger, *Comparaison d'intégrales orbitales pour des groupes p-adiques*, Proceedings of the International Congress of Mathematicians, Vol. 1, 2 (Zürich, 1994), Birkhäuser, Basel, 1995, pp. 807–816.

[72] _____, *Homogénéité de certaines distributions sur les groupes p-adiques*, Inst. Hautes Études Sci. Publ. Math. No. 81 (1995), pp. 25–72.

[73] _____, *Quelques resultats de finitude concernant les distributions invariantes sur les algèbres de Lie p-adiques*, preprint.

[74] _____, *Sur les germes de Shalika pour les groupes linéaires*, Math. Ann. **284** (1989), no. 2, pp. 199–221.

[75] _____, *Sur les intégrales orbitales tordues pour les groupes linéaires: un lemme fondamental*, Canad. J. Math. **43** (1991), no. 4, pp. 852–896.

[76] _____, *Une formule des traces locale pour les algèbres de Lie p-adiques*, J. Reine Angew. Math. **465** (1995), pp. 41–99.

List of Symbols

Ω, 1

Θ_π, 1

G', 1

q, 1

$\mathfrak{g}$, 1

$\mathcal{D}$, 1

J, 1

$J(\omega)$, 2

R, 2

T_L, 2

j_L, 2

B, 2

χ, 2

$\hat{f}$, 2

ℓ, 2

$\eta_\mathfrak{g}$, 2

$\mathfrak{g}'$, 2

$\mathcal{O}$, 2

$C_G(X_0)$, 2

$\mu_\mathcal{O}$, 3

$\mathcal{N}$, 3

$c_\mathcal{O}(T)$, 3

$c_\xi(\pi)$, 3

$c_0(\pi)$, 3

$d(\pi)$, 3

ϕ_f, 5

g_f, 6

f_P, 7

$\eta_{\mathfrak{g}/\mathfrak{m}}$, 8

$\eta_\mathfrak{m}$, 8

$w(A_2|A_1)$, 9

$A_1 \prec A_2$, 9

$A_1 \asymp A_2$, 9

$\mathfrak{g}_\mathfrak{h}$, 11

$^0\mathcal{D}$, 12

$d(\mathcal{O})$, 13

$r(\mathcal{O})$, 13

$C_\mathfrak{g}(X)$, 13

$\mathcal{O}(0)$, 13

$|X|$, 14

$\mathcal{O}(\gamma)$, 16

$\mathcal{D}_0$, 17

$\mu_\mathcal{O}(f)$, 17

$\mathcal{N}_d$, 19

$\mathfrak{z}$, 23

W_{ji}, 23

$[S]$, 25

J_0, 27

$\phi \sim 0$, 32

^cS, 32

R', 35

$\mathfrak{g}_e$, 38

ϖ, 42

S, 42

$\Gamma_\mathcal{O}$, 48

$\mathfrak{h}''$, 50

$\Psi_\mathcal{O}$, 51

$\Gamma_\mathcal{O}^\mathfrak{h}$, 51

$\xi_\mathfrak{g}$, 57

$\chi(A)$, 58

L^*, 58

$\mathfrak{g}(t)$, 58

f_X, 58

$J(V, t, L)$, 59

$C(V, t, L)$, 59

ϕ_α, 59

$J(V, \infty, L)$, 59

J_0, 59

$J_0(V, t, L)$, 59

$J_0(V, \infty, L)$, 59

$S(\mathfrak{g})$, 60

$I(\mathfrak{g})$, 60

$C(L)$, 62

$\mathcal{E}(K)$, 71

$\xi_{\underline{d}}$, 71

$d(\underline{d})$, 71

ξ_F, 71

$\mathcal{A}(F)$, 71

$\mathcal{A}(\underline{d})$, 71

$[\underline{d}:\delta]$, 71

$[F_1 : F_2]$, 71

$\mathcal{E}_{x,F_i}$, 73

1_{K_0}, 74

D_G, 75

$K^{1/2}$, 75

$\mathcal{E}^{1/2}$, 75

$\mathcal{O}_{\underline{d}}$, 76

U_M, 77

Θ_F, 78

$\mathfrak{m}_0$, 79

ϖ, 80

L_ν, 80

K_ν, 80

$K_\nu^{1/2}$, 80

Φ_ν, 80

$|T|$, 80

$\mathcal{O}_0$, 82

$\theta(\delta)$, 82

F_f, 86

$\mathcal{A}(\pi)$, 87

$m_\pi(L)$, 88

Index

admissible
  admissible distribution, 74
    (G, K_0)-admissible, 74
    G-admissible, 74
  admissible representation, ix
  admissible subset, 57
    $(G, \mathfrak{g})$-admissible G-domain, 58

Condition $C(L)$, 62
Condition $C(V, t, L)$, 59
cusp form, 12

distribution, x
  distribution character, x
  G-invariant distribution, 1

elliptic
  elliptic Cartan subalgebra, 5
  elliptic Cartan subgroup, 86
  elliptic orbit, 38
exp, 55

Fourier transform
  of a distribution, 2
  of a function, 2

G-domain, 3, 13
germ
  $\mathcal{O}$-germ, 48
  Shalika germ, 48

homogeneity
  of nilpotent orbital integrals, 18
  of Shalika germs, 48
Howe's Theorem, xii, 2, 62

interact, 71
intertwines, 71

lattice, 2
  adapted lattice, 58
    well adapted lattice, 58

dual lattice, 58
  s-lattice, 75
local character expansion, x
locally summable, 1
log, 55

nilpotent
  nilpotent element, 3
  nilpotent orbit, 3

orbit, 2
  dimension of, 13
  elliptic orbit, 38
  G-orbit, 13
  rank of, 13
  regular orbit, 30

regular
  regular element of $\mathfrak{g}$, xi, 2
  regular element of G, ix, 1
  regular orbit, 30

semisimple element, 3
smooth representation, ix

unipotent element, 3